Imagining Apocalyptic Politics in the Anthropocene

Bringing together scholars from English literature, geography, politics, the arts, environmental humanities and sociology, *Imagining Apocalyptic Politics in the Anthropocene* contributes to the emerging debate between bodies of thought first incepted by scholars such as Mouffe, Whyte, Kaplan, Hunt, Swyngedouw and Malm about how apocalyptic events, narratives and imaginaries interact with societal and individual agency historically and in the current political moment. Exploring their own empirical and philosophical contexts, the authors examine the forms of political acting found in apocalyptic imaginaries and reflect on what this means for contemporary society. By framing their arguments around either pre-apocalyptic, peri-apocalyptic or post-apocalyptic narratives and events, a timeline emerges throughout the volume which shows the different opportunities for political agency the anthropocenic subject can enact at the various stages of apocalyptic moments.

Featuring a number of creative interventions exclusively produced for the work from artists and fiction writers who engage with the themes of apocalypse, decline, catastrophe and disaster, this innovative book will be of great interest to students and scholars of the politics of climate change, the environmental humanities, literary criticism and eco-criticism.

Earl T. Harper is an Independent Scholar.

Doug Specht is a Senior Lecturer in the School of Media and Communication at the University of Westminster, UK.

Routledge Environmental Humanities

Series editors*:* **Scott Slovic** *(University of Idaho, USA),* **Joni Adamson** *(Arizona State University, USA) and* **Yuki Masami** *(Aoyama Gakuin University, Japan)*

The *Routledge Environmental Humanities* series is an original and inspiring venture recognising that today's world agricultural and water crises, ocean pollution and resource depletion, global warming from greenhouse gases, urban sprawl, overpopulation, food insecurity and environmental justice are all *crises of culture.*

The reality of understanding and finding adaptive solutions to our present and future environmental challenges has shifted the epicenter of environmental studies away from an exclusively scientific and technological framework to one that depends on the human-focused disciplines and ideas of the humanities and allied social sciences.

We thus welcome book proposals from all humanities and social sciences disciplines for an inclusive and interdisciplinary series. We favour manuscripts aimed at an international readership and written in a lively and accessible style. The readership comprises scholars and students from the humanities and social sciences and thoughtful readers concerned about the human dimensions of environmental change.

For more information about this series, please visit: www.routledge.com/Routledge-Environmental-Humanities/book-series/REH

Imagining Apocalyptic Politics in the Anthropocene

Edited by
Earl T. Harper and
Doug Specht

Routledge
Taylor & Francis Group
LONDON AND NEW YORK

First published 2022
by Routledge
2 Park Square, Milton Park, Abingdon, Oxon OX14 4RN

and by Routledge
605 Third Avenue, New York, NY 10158

Routledge is an imprint of the Taylor & Francis Group, an informa business

British Library Cataloguing-in-Publication Data
A catalogue record for this book is available from the British Library

Library of Congress Cataloging-in-Publication Data
Names: Harper, Earl T., editor. | Specht, Doug, editor.
Title: Imagining apocalyptic politics in the Anthropocene / edited by Earl T. Harper and Doug Specht.
Description: Abingdon, Oxon ; New York, NY : Routledge, 2022. | Series: Routledge environmental humanities | Includes bibliographical references and index. |
Identifiers: LCCN 2021015613 (print) | LCCN 2021015614 (ebook) | ISBN 9780367653095 (hardback) | ISBN 9780367653125 (paperback) | ISBN 9781003128854 (ebook)
Subjects: LCSH: Apocalypse in literature. | Apocalypse in mass media. | Politics in literature. | Human ecology and the humanities. | Fiction—History and criticism. | LCGFT: Literary criticism.
Classification: LCC PN56.A69 I43 2022 (print) | LCC PN56.A69 (ebook) | DDC 809/.933582—dc23
LC record available at https://lccn.loc.gov/2021015613
LC ebook record available at https://lccn.loc.gov/2021015614

ISBN: 978-0-367-65309-5 (hbk)
ISBN: 978-0-367-65312-5 (pbk)
ISBN: 978-1-003-12885-4 (ebk)

DOI: 10.4324/9781003128854

Typeset in Times New Roman
by codeMantra

Contents

Figures

Contributors

Martin Gren is an Associate Professor in Human Geography at the Linnaeus University, Sweden. He has a long-standing interest in the discipline theory of human geography, which has included works on the geographies of Torsten Hägerstrand and Gunnar Olsson. Martin's current focus is on the (re)conceptualization of the Earth in the context of the Anthropocene, the new climatic regime and the climate emergency. Martin is the author and editor of several books, including *Tourism and the Anthropocene*, the first volume that exposes tourism to the Anthropocene, and *GO: On the Geographies of Gunnar Olsson.*

Earl T. Harper is an Independent Scholar with no academic affiliation. He completed his PhD study on apocalyptic narratives of climate change and their influence on ecological gentrification in Western European cities at Bristol University in 2020. He is currently a Signaller with Network Rail and can often be found on the footplate of steam locomotives on the UK's mainline railway network.

Edward H. Huijbens (b. 1976) is a geographer and graduate of Durham University, England. He chairs Wageningen University's research group in cultural geography. Edward works on spatial theory, issues of regional development, landscape perceptions and the role of transport in tourism and polar tourism. He has authored over 40 articles in several scholarly journals and edited volumes, published three monographs in both Iceland and internationally, and has co-edited four books.

Philip Jones received his PhD in 2018 for his thesis "Rewriting the Atlantic Archipelago: Modern British Poetry at the Coast." He has written on power, gender and watery embodiment in the work of Wendy Mulford, Frances Presley and Carol Watts, as well as on representations of atomic agency in the poetry of Robert Minhinnick. He has taught at Oxford Brookes University and the University of Nottingham.

Charlotte Lancaster is a PhD Candidate in the Environmental Humanities at Bath Spa University. Her research focuses on representations of flooding and other "natural disasters" in Anglophone fiction from the nineteenth century to the present. Charlotte's dissertation examines a variety of novels that revisit the mythic flood tradition as an interpretive framework to communicate the philosophical and ethical implications of climate and environmental crisis.

Aapo Lunden is a PhD Researcher at the Geography Research Unit, Tourism Geographies Group, University of Oulu, Finland. His research focuses on critical sustainability studies, institutions and governance of land use and tourism in protected areas in the Arctic.

Nicholas Ferris Lustig is an Assistant Professor in the Department of Geography at the University at Buffalo. His primary research project analyses the emergence, spread and opposition to the mass surveillance programmes and real-time crime centres constructed by police departments throughout the USA.

Carl Christian Olsson is a Writer and Doctoral Student at Newcastle University. His current research centres on cashing out attempts to naturalize Kant's epistemology in relation to the history of geographical thought and attendant terrestrial discourses in philosophy and literature. Carl is also a Certificate Student and researcher at The New Centre for Research & Practice and holds an MSc in Human Geography from Lund University.

Diana Barquero Pérez is a Costa Rican graduate of the Master of Arts in Spatial Strategies at Kunsthochschule Weißensee (KHB) Berlin (2020). She is also Bachelor of Fine Arts Major in painting at the University of Costa Rica (2014). Her artistic work visualizes processes of friction, transformation and collapse in peripheral spaces. The artist combines theoretical research, fieldwork and material exploration to produce artistic research linked to political ecology, critical geographies and landscape studies. As part of her professional work, she engages in collaborative processes with professionals from other fields such as environmental sciences.

Lucas Pohl is a Postdoctoral Researcher at the Department of Geography of Humboldt-University Berlin, Germany. He received his PhD at the Department of Human Geography of Goethe-University in Frankfurt with a dissertation thesis that elaborates on a psychoanalytic approach to urban ruination. More generally, he works on the interstices between philosophy, psychoanalysis and urban geography, focusing on social and spatial theory, built environments and political action. The latest results of his research have been published in *Cultural Geographies*, *Social & Cultural Geography*, *Housing Studies* and *Theory, Culture & Society*.

Mariana Reyes-Carranza is a PhD Candidate at the School of Geography, Queen Mary, University of London. She hold a BS degree in Biology from the University of Guadalajara and an MSc in Environment, Culture and Society from the University of Edinburgh. Her research interests lie at the intersection of political ecology, critical pedagogies and environmental justice.

Cat Snyder is a communications professional with experience across human rights, healthcare and housing. Born in the USA and raised in Hungary, Cat gained a Bachelor's degree in Intercultural Studies in California. With an MA in Diversity in the Media from Westminster University in London, her focus has remained in advocacy particularly within vulnerable populations. Her research focuses on political and social change amongst migrant and other marginalized communities.

Doug Specht is a Chartered Geographer and a Senior Lecturer in the School of Media and Communication at the University of Westminster. His research examines how knowledge is constructed and codified through digital and cartographic artefacts, focusing on development issues in Latin America and sub-Saharan Africa, where he has carried out extensive fieldwork. He has published widely on data ethics, development and humanitarian response.

Tristan Sturm is a Senior Lecturer in Geography at Queen's University Belfast. He is interested in apocalyptic thought and conspiracy theories related to climate change, geopolitics and religious movements. He has published on these themes in the *Toronto Star*, *Haaretz*, *Jerusalem Post*, *CounterPunch*, *Journal.ie*, *National Post* and *The Magazine*. He is co-editor with Jason Dittmer of *Mapping the End Times*.

Samo Tomšič holds a PhD in Philosophy from the University of Ljubljana, Slovenia. He is currently a Research Fellow at the Humboldt University, Berlin and a visiting researcher at the University of Ljubljana. His research areas comprise political philosophy, structuralism, psychoanalysis and epistemology. His recent publications include *The Capitalist Unconscious: Marx and Lacan* (Verso, 2015) and *The Labour of Enjoyment: Toward a Critique of Libidinal Economy* (August Verlag, 2019).

Carlos Tornel is a PhD Candidate at the Department of Geography, Durham University, UK. His research focuses on energy landscapes, decolonial forms of environmental, climate and energy justice and environmental politics.

Charles Travis is an Assistant Professor of Geography and GIS with the Department of History at the University of Texas, Arlington, and a Visiting Research Fellow with the School of Histories and Humanities at Trinity College Dublin. As an editorial board member of the journal *Literary Geography*, his book publications include *Abstract Machine: Humanities GIS* (2015), *History and GIS: Epistemologies, Reflections and Considerations* (2012) and *Literary Landscapes: Geographies of Irish Stories, 1929–1946* (2009). His work has also appeared in *Annals of the Association of American Geographers*, *Global and Planetary Change*, *International*

Journal of Humanities and Arts Computing, International Journal of Geographical Information Science, Historical Geography and other peer-reviewed publications.

James A. Tyner is Professor of Geography at Kent State University and a Fellow of the American Association of Geographers. He is the author of over 20 books, including *War, Violence, and Population: Making the Body Count*, which received the AAG Meridian Book Award for Outstanding Scholarly Work in Geography.

María Soledad Castro Vargas' research is focused on analysing the relationship between chemicals and socio-natures, combining approaches from political ecology, critical geographies and ecotoxicology studies. She is interested in chemical geographies of pesticides and contaminants of emerging concern and waterscapes studies. Her doctoral thesis is grounded in the case of the pesticide-contaminated wetland Térraba-Sierpe in Costa Rica and it aims to understand why and how contaminated waterscapes are being produced in Costa Rica in relation to the pesticide complex. She holds a BSc in Natural Resources Management and an MSc in Water and Coastal Management.

Stephanie Wakefield is an urban geographer whose work critically analyses the technical, political and philosophical transformations of urban life in the Anthropocene. She is currently Director and Assistant Professor of Human Ecology at Life University. She is the author of *Anthropocene Back Loop: Experimentation in Unsafe Operating Space* and co-editor of *Resilience in the Anthropocene: Governance and Politics at the End of the World.* Her new book, *Urbanization in the Anthropocene*, analyses experimental sea rise adaptations in Miami and traces an emergent "urbicidal Anthropocene" that challenges both urban resilience and planetary urbanization.

Introduction

… these unprecedented times

Earl T. Harper and Doug Specht

Everyday the world gets closer to ending. Everywhere, the harbingers of the apocalypse appear: our water, air and soil are polluted and dying; our own bodies harbour the micro-plastics that seemingly every commodity now sheds; new viruses sweep across the globe, changing the way we relate to one another and our daily work; new mines and oil pipelines are approved as climate and environmental scientists watch helplessly from the side-lines; the bees are going extinct and taking most floral and fruiting life with them; acoustic pollution of the oceans is driving whales and dolphins to beach themselves; and the rise of fascist and right-wing politics across Europe and North America seems to bring no promise of relief from the economic and political shit we find ourselves in. However, there is some hope amongst the disaster and catastrophe of global geopolitics: as mutual aid organisations step up to the plate and provide some of the essential services that keep communities going during crises such as pandemics and hurricanes; massive mobilisations of school children and ecological activists take to the streets and strike for our collective futures; new forms of economics, such as so-called 'doughnut economics', are becoming a matter of policy for cities like Amsterdam; social media provides important methods for continuing social interaction whilst being physically isolated; and through astounding amounts of public funding and international cooperation, the process of developing a vaccine for COVID-19, something which would normally take over a decade, was achieved in just nine months. This book is as much about understanding the apocalyptic current as it is about understanding these collective and individual responses to them. To understand the apocalypse, however, an unlikely source of inspiration is turned to: fiction and narrative.

Accompanying this rise in 'real' world (a term which will become more problematic as this book continues), apocalypses is also a corpus of literature, films, artwork, academic writing and popular narratives which imagine the various ways in which the world might end. Beginning[1] with books such as Aldous Huxley's 1932 *Brave New World* and the 1954 Richard Matheson novel *I am Legend* and continuing through to Olafur Eliasson's 2018 installation at the Tate Modern named *Ice Watch* and blockbuster films like *The Day After Tomorrow* (Emmerich, 2004) and any one of a number

DOI: 10.4324/9781003128854-1

of Neill Blomkamp films made in the early 2000s and 2010s, Western society seems to be fascinated by its own demise. Films are a particularly pernicious and pervasive media for this, be it through zombies [*Walking Dead* (Fox Broadcasting Company, 2010), *I am Legend* (Lawrence, 2007), *Shaun of the Dead* (Wright, 2004), etc.], plagues [*28 Days Later* (Boyle, 2002)], biological demise [*Children of Men* (Cuarón, 2006), *Logan's Run* (Anderson, 1976)], climate change [*The Day After Tomorrow* (Emmerich, 2004), *Snowpiercer* (Joon-ho, 2013), *Flood* (Mitchell, 2007)], nuclear accident or war [*Dr. Strangelove* (Kubrick, 1964), *The China Syndrome* (Bridges, 1979)], ancient prophecy [*2012* (Emmerich, 2009)] or even climate change induced, teleporting, shark-infested, tornadoes (*Sharknado 1–5*), narratives of total apocalyptic destruction of Los Angeles, New York, London, Paris, Shanghai and the world in general seem to bubble behind the mainstream of Western cinema (we choose to focus on cinema here because of the sheer scale of capital flows which accompany it, making it one of the most highly valued narrative fields in contemporary, Western society). It is interesting to note that many of the existential threats faced by society are mirrored in one way or another by these films (Dixon, 2003; Hall, 2011; Hantke, 2010; Platts, 2013). For example, the long-felt threat of invasion of the United States by the Soviets in the 1950s and 1960s was represented by a plethora of alien invasion films, whilst the relentless march of unbridled neoliberalism in the early 2000s was accompanied by a renaissance in the zombie genre. During the beginning of the COVID-19 pandemic in early 2020, the most-watched films on popular streaming service *Netflix* were titles such as *Contagion* (Soderbergh, 2011), *Outbreak* (Petersen, 1995) and *28 Days Later* indicating a collective desire to revisit narrative tools to understand our collective position within the apocalypse which appeared to be emerging in China, Europe and North America.

Indeed, one of the challenges in completing this manuscript has been the incredible rate at which apocalyptic events have emerged during the time of writing (2019–2021). The idea for the book emerged in 2019, when climate change was beginning to gain traction in the public imaginary, through the *#Fridays4Future* twitter movement, and simultaneously, the rise of fascist politics in the United States seemed assured after three years of President Donald J. Trump, the proud boys, and tiki torch protests in Charlottesville. The United Kingdom was beginning to understand the full and disastrous consequences of having voted to leave the European Union, whilst the third general election in two years had put Boris Johnson – a man widely known for his racist and homophobic buffoonery – at the helm of one of the most difficult and historic moments in recent British history. As the flow of refugees and asylum seekers from Honduras and El Salvador arriving in the United States increased, ex-President Trump's administration began detaining them and their children in 'cages' on the US-Mexico border. Tensions in the Persian Gulf over shipping and oil access and drone activity by the United States and Iran made the likelihood of another Gulf War extremely likely whilst large swathes of the Brazilian Amazon rainforest we set on fire, severely damaging the ancient ecosystem in the name of agricultural progress. Hindu Nationalism in India revoked Kashmiri autonomy, reigniting tensions in the region leading to curfews, blackouts and mass arrests of Muslims and the end to secular state of India. President Trump was impeached for putting pressure on Ukraine to investigate his opponents in the upcoming elections. Hong Kong erupted in protest, as did Algeria leading to the resignation of President Abdelaziz Bouteflika. Sudan followed in protest which pushed out President Omar Hassan al-Bashir, whilst a public transport fee hike meant that Chileans also took to the streets. Protests in Lebanon, Iraq, Bolivia, India, Nicaragua and Russia erupted defining 2019 by disaster, state violence and protest.

When work began on the book, the Australian bush was alight, followed not long after by the Californian forests, blacking out the sky for months and making firefighters pray for rains that were delayed by months due to climate change. Tensions between the United States and North Korea created the conditions for the threat of nuclear war to re-emerge and then, in February and March, reports of a new virus in Wuhan, China, began to circulate. The COVID-19 virus then spread westwards at an ever-increasing pace, locking down countries in its wake. The United States took the opportunity to name it the 'China virus' in response to rumours that the Chinese government had known about and covered up the viral outbreak as early as November 2019, another tool in the ongoing trade wars between the two countries and covering up the UK and US government's own ineptitude and inaction. On 25 March 2020 an event which forced the European, American and white colonialist superpowers to examine their own systemic racism took place: the murder of George Floyd by Minneapolis police officer Derek Chauvin for having allegedly used a counterfeit $20 bill to buy cigarettes. The death, in itself, was the most recent in a long line of deaths of African Americans at the hands, muzzles and knees of white police officers: Breonna Taylor, Michael Brown, Atatiana Jefferson, Eric Garner, Aura Rosser, Philando Castile, Michelle Cusseaux, Oscar Grant III, Janisha Fonville, Walter Scott, Terence Crutcher, Daniel Prude, Stephon Clark and Botham Jean are just some of the names of African Americans killed for falling asleep in their cars, holding mobile phones, eating ice cream, being depressed, or for informing officers that they had a legal firearm in their car at a routine traffic stop. The death of George Floyd, however, due to the public, brutal execution of an officer kneeling on his neck in the street for close to nine minutes, even after the paramedics arrived, galvanised a movement. The Black Lives Matter protests erupted and led to a global wave of recognition for the need to decolonise every aspect of public life. Towards the end of 2020, just as hope began to emerge with election of President Joe Biden Jr and the first black, Asian, female Vice President, Kamala Harris, in the United States and the release of a vaccine against a virus that has killed 2 million people globally at the time of writing, new, more transmissible variants of the virus emerged in London, then South Africa and Brazil. In the early days of 2021 as we finish work on the book, an attempted coup has been staged to keep former President Donald Trump in power, the UK government has distributed food packages to hungry children which are not fit for purpose, whilst the leader of the House of Commons, the Right Honourable Jacob Rees-Mogg has called it 'disgraceful' that the United Nations (UN) has stepped in to feed children in the United Kingdom for the first time in history. Not 'disgraceful' because there is a need to do so, but 'disgraceful' because he believes the UN is playing politics to make his government look bad.

To be sure, we live in unprecedented times. The realisation, of course, being that our species and our planet may have arrived in the apocalypse long ago. Welcome back, to the apocalypse.

Welcome to the apocalypse, it's nice to see you again

In general, the apocalypse is taken to mean the complete and total destruction of the world. However, the various apocalypses we tell ourselves we are in are not solely defined by 'the end' and are as much about finding new beginnings. For many years, a debate has taken place in the field of eschatology as to whether an apocalyptic event is as simple as the Hollywood films we watch would have us believe. For most, the notion of the apocalypse conjures a sense of ending, of total destruction and of sadness. One only has to look at how often Los Angeles has been destroyed on the silver screen to understand that the 'apocalyptic' genre of film takes total destruction as its starting point (Dixon, 2003). However, if we begin to look closer at the apocalypse, we can witness several key tropes at play. The first of which is that, yes, something is destroyed (Svoboda, 2016). But, as so many artists (as well as those who work with semiotics and communication) would remind us, the thing which is destroyed in films often stands in for something else. In *Independence Day* (Emmerich, 1996), when riotous applause spontaneously erupted in audiences in the United States as the model of the White House was destroyed (Troxler and Pearsall, 1996), they were perhaps not really cheering the destruction of the building, but the system of government which it stood in for in the film. A brief reminder that, in the face of the apocalypse, nobody is immune, not even the powerful, political elites of Washington, DC. So, if destruction of objects becomes destruction of symbols, what can we learn from the symbolic apocalypse?

The second trope of apocalyptic narrative is often that, after the cataclysmic event of the story, a form of 'normality' begins to resume and reassert itself – balance is restored to the world and life can once again move forward (Svoboda, 2016). To understand this, the etymology of 'apocalypse' is useful to revisit. Importantly, the root of the word 'apocalypse' comes from the ancient Greek term *αποκαλύπτειν* [*apokalýptein*] which translates roughly as 'unveiling' or 'revealing'. The conflation of apocalypse with catastrophe

comes, in part, from the recognition that for the unveiling to happen, a transition from the 'old' world to the 'new' world takes places – the destruction of the old leads to the new in an act of revelation. The question becomes, then, what is the revelation of the contemporary era and what could or does the 'new' look like? If we are to describe current affairs as 'apocalyptic' then there must be some basic understanding that they can reveal important things that could lead to the destruction of the old ways and the construction of new ways, surely?

One of the key principles to recognise in this practice of narrating the apocalypse is that the total destruction of the planet, or our way of life, or our homes – whatever scale the narrative functions at – is a depressing story to both tell and receive. The idea that 'revelation' comes from that destruction provides a way for humans to be hopeful that a true apocalyptic end of the world might have a 'Eureka!' moment of discovery and salvation, too. Back in 2013, a workshop was held at the University of Manchester in which Erik Swyngedouw, Neil Smith and Andy Merrifield had gathered to discuss the ongoing revolutions and eruptions of democratic politics of *Occupy!*, the Arab Spring and student tuition fee protests. The general narrative of that meeting was one of hope: perhaps this will be the moment that we overthrow capitalism? Pre-figurative politics are finally being enacted after our years of proposing them as a new way forward! What will happen to the role of the critical academic if we no longer have anything to criticise? Of course, these three scholars were attempting to make a catastrophe less *catastrophic* and more *apocalyptic* – to glimpse what might be learned from the financial crisis, to carve out an apparently inevitable new world in the potential death of the old. Unfortunately, the old world gathered strength and capitalism and right-wing politics quickly swept away the scholarly community's apocalyptic optimism. However, perhaps the true power of this moment is witnessed in the continuation of the tactics of the movements of the squares in contemporary acts of resistance – the revelation gleaned from earlier movements was of effective ways to resist the coming doom, and those methods of resistance live on in Black Lives Matter and Extinction Rebellion activists.

It is also of fundamental importance that we recognise the colonial tendency to think of the apocalypse as a future event, when, especially for many indigenous cultures, it has already happened at least once (Davis and Todd, 2017; Whyte, 2018). As Carranza (this volume) discusses, the arrival of settler colonialism in many South American Indigenous cultures is signified through apocalyptic tropes of toxicity, destruction, disease and decimation of landscapes, peoples and the sky itself. This sentiment is also argued by Kyle P. Whyte (2018) who argues for decolonising the Anthropocene by incorporating an understanding of climate change and environmental degradation which is already informed by previous experiences of colonial violence and epistemicide conducted against Indigenous peoples. As a result of the dominance and ease of access of Euro-American narratives of the apocalypse, these cosmologies and histories have traditionally been underplayed or excluded from debates around the Anthropocene and the apocalypse – in this book, however, every attempt has been made to explore these ideas in the chapters contained within.

The apocalyptic politics of the Anthropocene

This book is an attempt to explore these sorts of questions, and if we are so bold as to suggest that this is possible, offer some potential lessons which can be learned from the multiple apocalypses that these chapters discuss. The fundamental question the authors engage with in this volume is simply this: does the apocalypse (in whatever form it may take) give rise to new politics or entrench old forms?

For example, early works by Slavoj Žižek (1999, 2009) suggest that the apocalyptic event provides the opportunity for the implementation of new ideas and ideologies which might otherwise be unpalatable. Here, Žižek is of course drawing on and complementing the brilliant work of Naomi Klein who, in *The Shock Doctrine* (2007), argues that military shock and awe tactics bewilder and traumatise subjects into a sort of *tabula rasa*, a blank slate, upon which new ideas and practices can be imprinted. For Klein, this explains the success of such tactics in US military interventions in implementing capitalism and installing new governance regimes in countries in South America and the Middle East. Žižek draws on her work and suggests that the solutions to environmental catastrophe open up new channels of investment, allowing the system to commodify the solution to the problem, even if that problem was caused by the system itself. In this way, for Žižek, the environmental, political or economic catastrophes that seem so common in the modern era are essential for providing much-needed avenues of investment and capital circulation.

More recently, the apocalypse has been theorised by Erik Swyngedouw (2010, 2013) as a depoliticising force. Following Badiou's (2008) assertion that 'Ecology is the new opium of the masses' Swyngedouw argues that the relationship between a scientific foundation for acting and the proclivity of capitalism to produce new solutions to the problem is, in itself, problematic. The reference to ecological imperative – 'buy this or else we will all die' – provides otherwise debatable and questionable approaches to the crisis with an incontestable force for implementation.

Interestingly, with the rise of popular politics and movements such as *Indignados*, the *15M* movement, the *Umbrella Revolution*, *Occupy!*, and the increasingly visible profile of the *Zapatistas*, Žižek's claim that the apocalypse's only power is to foreclose alternatives changed. With the hopeful politics practised after the 2008 global financial crisis, Žižek (2011) argued that the apocalypse is precisely a moment where the dominant structures can be seen for what they are, helping to dismantle and reconfigure society for the better. *Occupy!*, then, was not a spontaneous eruption of a new form of politics but a reaction to the apocalyptic revelation that financial capital could cease to flow and the dominant structures of politics and economics would focus on saving those who caused the crisis rather than those affected by its cruelty. This profound change in the implications of the apocalyptic moment does not seem to have been sustained in the way Žižek thought it might, but it does seem to have changed politics in quite fundamental ways, lending support to the concept.

Alongside these discussions of the role of apocalyptic moments and crisis in influencing our ability to form new ideas and ways of doing is the emergence of the Anthropocene as a key concept in social sciences and the humanities. Much has been written about the concept, and we do not wish to detract from those important and eloquent debates in the limited space we have for this introduction, but we are at an interesting moment in our collective understanding of the world and our place in it. Slowly but surely, human civilisation has come to place itself in a position where it can have global impact on the planetary ecosystem (Bonneuil and Fressoz, 2016; Clark, 2015; Zalasiewicz et al., 2011). This knowledge can go either of two ways: we realise that we have dominance over the planet and have already fundamentally changed it for the worse or we begin to function as careful custodians of the planet, able to reshape it into a healthy cohabited environment until such a time as our influence diminishes. The knowledge of our place in the ecosystem makes action important, but considered action even more important. For this reason, the concepts discussed around apocalyptic politics are essential for understanding which way our future might lead us.

Narrative desires and desiring narratives

The question then arises: why look at fiction, art, performance, social media and narrative rather than actual, really existing apocalyptic scenarios? To answer this question, the field of psychoanalysis is illuminating. Since Christian Metz wrote his foundational text, *Psychoanalysis and Cinema: The Imaginary Signifier* (1982), many scholarly attempts have been made at relating the desire for narrative to our ability to function as subjects in the contemporary world. Importantly, Lebeau (2001) argues that film (and here we could read any form of narrative device: books, artworks, etc.) acts as a 'collective dreamstate'. For Freud, the dreamstate of an individual could be read, revealing repressed desires and fears (Shortland, 1987), so film, in this instance, could be read as revealing the repressed desires and fears of the societies which consume and produce them. In short, the stories we tell each other reveal what we would really like to do, should we not be constrained by society. Narrative is, as Metz reminds us, *The Imaginary Signifier*, signifying our collective imagination and shaping how we can imagine.

This last point is important, because not only does film reflect our desires, but it also shapes and forms what it is possible to desire in the first place. The Mirror Stage in Lacanian psychoanalysis, when a child of around 18 months of age begins to learn about their own body by witnessing their reflection in a mirror (Lacan, 2006), arguably also functions through any form of screen – including cinema. We tend to prefer the stories which contain characters or scenarios which we can identify with: maybe the main character is sensitive, caring and an eco-warrior; perhaps the protagonist finds themselves battling the corporate machinery to get the 'truth' out into the world; the beautiful woman ends up with the beautiful man in the end. This need for identification and reflection of our own world in the world's we read and watch through stories is in part because it helps us to deal with future situations which we do not yet have the tools to cope with (Lacan, 2006; Lebeau, 2001). Quite literally, we come to identify with the characters and situations as we form our identity around watching them. This is precisely why studying apocalyptic narratives in a really existing apocalyptic world is of such importance, because it allows us to glimpse how our societies deal and might deal with future events. The observation that most, if not all, apocalyptic cinema ends with a 'return to normal' becomes particularly poignant here. As Žižek states 'Cinema is the ultimate pervert art, it doesn't *give* you what you desire, it *tells* you how to desire' (Fiennes, 2006).

The role of apocalyptic narrative has been written about in various ways before, for example the brilliant work of Evan Calder Williams (2011), laying the foundations for understanding the current moment as a combined and unevenly distributed apocalypse, and offering ways of understanding the future through steampunk and other storytelling cultures. Timothy Clark (2015) has highlighted the importance of understanding the Anthropocene through ecocriticism and narrative devices. Giuseppina Botta (2010) explores the literary relationship between Margaret Atwood and apocalyptic bioengineering of our own internal nature in the form of DNA manipulation. Zumbansen and Fromme's (2010) work explores to what extent ecocritical elements have permeated through film and text with the growing realisation that the current crisis is a human-caused one, whilst E. Ann Kaplan (2016) has defined a new form of psychological condition, pre-trauma, based on the role of film and narrative in shaping our fears and inability to act on the future because of trauma yet to occur. The growing recognition of the importance of these ways of narrating imagined or real apocalyptic events for shaping our ability to imagine and desire various kinds of future gives this book its starting point.

So, what does this book aim to offer to the quagmire of global catastrophe? The concept emerged in a discussion between like-minded researchers engaging with similar topics at the 2019 Royal Geographical Society Annual Conference. The clear and obvious message of the papers presented during a session organised by one of this book's editors, Earl Harper, along with Jonathon Turnbull, was that the collective societal imaginary – particularly in the global North – is alive with apocalyptic stories, narratives and perceptions of the current moment. At the point of organising the session in early 2019, it seemed the main apocalyptic events about to happen were mainly articulated around climate change and micro-plastic pollution but the session raised questions about Western-centric ideas of the apocalypse, anti-microbial resistance, urban ecological soundscape destruction and several other biotechnical apocalypses. One of the strong themes which developed through the session was the importance of narrative and storytelling in helping to understand how the apocalypse might play out and how we, as a species, might respond to those narratives. This book is, in part, an attempt to explore those issues. As we write this introduction, in early 2021 the SARS-CoV-2 virus has stalled global economies, set nations against one-another and called into question the very foundations of how we 'value' different forms of labour in the West. Importantly, this apocalypse is precisely an 'apocalypse' rather than a disaster or catastrophe because it unveils previously hidden or ignored relationships.

To explore these ideas further the authors of this volume take us on a journey through literature, film, the smartphone, and our own minds, revealing what it is to live wrapped within the apocalyptic politics of the Anthropocene.

In their opening salvo **Edward H. Huijbens** and **Martin G. Gren** lay out the issue that confronts us. That is that 'our house is on fire'. The planet is in dire straits, with runaway warming already biting at our heels. The authors of Chapter 1 note that the hottest days on record have occurred during the writing of this volume, and that this is just the beginning – parts of the world previously inhabited are already becoming too hot for humans to occupy. The Amazon rainforest nears a tipping point at which its destruction becomes irreversible, and the Greenland ice sheets are melting faster than predicted. These apocalyptic changes have not gone unnoticed, and around the world people have declared a climate emergency, but the authors also note that this casts what they call a 'murky spell' over the population, leading us towards 'existential, emotional and cognitive dissonance' in relation to our burning house. Drawing upon this emotional crisis Chapter 1 leads us into understanding why the concept of the Anthropocene is so important to understanding our present position as guests on planet Earth. The problem though is not so much how we got here, but how do we go on living in a house that is on fire?

In Chapter 2 **Carlos Tornel** and **Aapo Lunden** urge us to not yet *go gently into the good night*, but rather that the important work is in understanding our imaginations of the apocalypse. The contention though is not as to whether the threats to existence are real, but about how we respond to them. Positioning the apocalypse as *Other* does not lead to change, but to resignation and this must be avoided at all cost the authors note. Invoking the apocalypse is indeed self-fulfilling, leading to inappropriate, piecemeal development and market-based innovations to address the Anthropocene. This chapter then foregrounds the coming book, suggesting as it does that the true challenge of the Anthropocene is to win the war of the imaginary, and to identify inequalities in dreams of utopias and dystopia that lead to the narrowing of alternatives to human demise.

Chapter 3 then builds upon this notion, opening with the worlds 'another end of the world is possible', and going on to challenge just such inequalities. Here **Mariana Reyes-Carranza** continues to draw out the internalised inevitability of the end of existence, drawing upon indigenous cosmologies of climate change. Taking the mythical universe of the Yanomami people this chapter plots the end of the world by re-examining *The Falling Sky*, a text co-authored by shaman Davi Kopenawa Yanomami (2013) and anthropologist Bruce Albert. Reyes-Carranza further employs an analysis of the doomsday imaginaries presented in the *Museum of Tomorrow*, Rio de Janeiro, Brazil. In bringing the two sites together the chapter questions and contrasts narratives of the Anthropocene that reinforce or reject human mastery over the planet. Many explorations, and indeed explanations, are entangled within precisely the conditions which led us to this brink argues Reyes-Carranaz. Instead, she suggests that indigenous cosmologies offer a powerful form of eco-political resistance in this time of acute emergency.

Charlotte Lancaster uses Chapter 4 to also explore the relationships between indigenous peoples and the apocalypse. Through a re-reading of George Turners's (1987) *The Sea and Summer*, Lancaster challenges the absence of Australia's Aboriginal Indigenous communities. Yet, at the same time she notes that the narrative of the text, through its repeated apocalyptic events, echoes the sustained structural violence endured by the traditional inhabitants of Turner's fictionalised vision of Melbourne's Dandenong Ranges. Lancaster forces us to look again at the conspicuously white narratives of apocalyptic imaginations, and their relation to settler-colonial capitalism, and to ask what histories are denied in these tales.

In Chapter 5, **Charles Travis** also looks to stories and literature in his exploration of our imagined demise. Drawing upon *The Tempest* (1611), *Frankenstein* (1818) and *Cloud Atlas* (2004). Using the work of Hannah Arndt as his basis, Travis leads us towards the idea that these apocalyptic imaginaries are both the construct and constructor of the Anthropocene. These three works, along with Byron's (1816) *Darkness*, could well be seen as the true indicator of the much-contested start of the Anthropocene, documenting and predicting the Armageddon through the eyes of Prospero, Frankenstein's nameless monster, Zachry Bailey, and the dwindling stocks of Atlantic Cod. This chapter takes us deeper into the world of apocalyptic literary geographies, leading us to reimagine familiar stories as world ending narratives.

Chapter 6 continues the exploration of literature as **Philip Jones** tackles the concept of Geotrauma and the work of mourning in Jorie Graham's (2019) *Fast*. The poetry of Graham speeds us towards the time of final arrival, but their experimental rhythms interrupt these flows, pulsing and forcing us to confront the inevitable systemcide that approaches. Jones takes us through these ebbs and flows, both drawing out the ecological peril within Fast, and at the same time providing a comforting companion he moves the reader towards a collective mourning for a world that is increasingly untenable as a site for dwelling. This is a painful journey, but it is an important one, for as Jones quotes, 'when loss remains unspoken, neither grieved nor worked through, then change and adjustment cannot follow' (Randall, 2009).

The comforting companionship offered by Jones is much needed as Chapter 7 unfolds. **James A. Tyner** confronts us anew, asking 'what is a meaningful life at the end of times?' Taking William Nolan and George Johnson's (1967) dystopian classic, *Logans Run* as the jumping off point for this provocation, Tyner highlights the often-disturbing parallels between the text and the contemporary society. Through the chapter's waving together the work of Nolan and Johnson with modern-day politics and thinking, *Logans Run* ceases to be an apocalyptic imagination and becomes a lived reality. Most disturbing of all, it is clear that the fictionised *duty to die* is as much a reality as anything. The value, and meaning, placed on life is brought into sharp focus by Tyner's analysis of coronavirus, where people were all too readily informed it was their duty to not be treated, to free up beds, to make way for youth. In the Anthropocene then, is a meaningful life one that ends for the good of continued existence of the planet?

Further questions around apocalyptic narratives of the Anthropocene in film and literature are examined by **Lucas Pohl** and **Samo Tomšič** in Chapter 8. Through an exploration of Freudian and Marxist thought, and borrowing from the works of Lacan, they note how apocalyptic imaginations are presented as wastelands of asociality, in which the bonds between humans no longer exist. This, Pohl and Tomšič argue, hides a darker truth, one that is more difficult to confront and so shied away from. That is that the apocalypse is an inherent, unavoidable part of the capitalist ordering of the social. The Anthropocene is then epitomised as both the dominance and redundancy of the human. The question of where the human is placed within the Anthropocene is challenged, and with it our own imaginations of futures.

Against this backdrop of human redundancy **Doug Specht** and **Cat Snyder** explore the self(ie) in the Anthropocene in Chapter 9. In asking what drives us to take selfies whilst protesting, marching and fighting against climate change, this chapter opens up questions on what it is to exist, how we represent ourselves and why producing photos of ourselves is not the narcissistic act assumed by many, but rather one that is essential. Specht and Snyder argue that these digital portraits, shared across the internet, become necessary because as the apocalyptic imagination grips us more firmly our sense of perpetual existence diminishes. The selfie then is all that remains to record our fleeting moments on a dying planet. In this context selfies in the Anthropocene are both an individual and collective ritual of mourning for ourselves and the environment.

In Chapter 10, **Stephanie Wakefield** turns our attention to the city. Urbanisation she notes is often cited as the driving force behind the Anthropocene. Yet these cities are also the sites of trialling climate adaptation – living laboratories that seek to find mitigation for the harms caused by their own construction. Examining Miami, Wakefield unearths the contradictions between those actions that seek to preserve the city and those that may destroy it. The concept of the end of the city is not new she notes, it is after all the city that plays a central role in apocalyptic imaginations, but the demise of the city manifests in diverse ways. Creating new imaginations is important to the remodelling and re-understanding of Miami, and is required to halt, or at least slow urbancide.

From the macro we turn to the micro in Chapter 11 as **María Soledad Castro Vargas** and **Diana Barquero Pérez** take us into the world of micro-plastics. These imperceptible flakes of plastic are polluting the world around us, rendering water deadly for creatures that live within, as well as those who are forced to drink it. Whilst these chemicals serge around our own bodies, Castro and Barquero through their haunting photography force us to look again at the contradiction between the enhancements in health and productivity provided by the very plastics that now poison us and the planet around us. These contaminants have become a fetish of capitalism, and our own demise.

As we draw towards the close of the book, **Carl Olsson** challenges us anew in Chapter 12. In a reimagining of Deleuze and Guattari's (1987) *A Thousand Plateaus*, Olsson twists our own memories of this work and our collective imaginations as he pulls us towards the creation of new memories of the apocalypse. Remaining a pure as possible to Brian Massumi's translation of the text, and for the most part relying on sources that would have been available to Deleuze and Guattari, Olsson builds and expands their thesis through his experimental apocrypha opening new avenues for the imaginations of the Anthropocene and apocalyptic politics. If death is inevitable, the question is then: under what circumstances can one afford to be pocked with the requisite caution?

In the final chapter of the volume, **Tristan Sturm** and **Nicholas Lustig** return us to the present, and our less imagined, but imminently unfolding ecological breakdown. The apocalyptic imagination, the future-orientated and futurist thinking that grips us tightly is also what makes up impotent to change. It is the work of apocalyptic discourses to promote and maintain the post-political condition they argue. Imaginations can do little for us; instead, we need immediate policy and political intervention, we need the emancipation of the people, we must look to the discourses and contestations of the now, not just some imagined future. To this end, Sturm and Lustig conclude their chapter, and thus this book, with six core principles of a new radical apocalyptic discourse.

Note

1 The trope of apocalyptic storytelling, of course, begins much earlier than 1932, but these works are often considered the 'beginning' of the current trend and are sufficiently well known currently to be included here as significant.

References

Anderson, M. (1976) *Logan's Run*. [Film] United States: Metro-Goldwyn-Mayer.

Badiou, A. (2008) Live Badiou – Interview with Alain Badiou, Paris, December 2007. *In:* O. Feltham ed. *Alain Badiou: Live Theory*. London: Continuum, 136–139.

Bonneuil, C. & Fressoz, J.-B. (2016) *The Shock of the Anthropocene: The Earth, History and Us*. London and New York: Verso.

Botta, G. (2010) Faustian Dreams and Apocalypse in Margaret Atwood's *Oryx and Crake*. *In:* L. Volkmann, N. Grimm, I. Detmers & K. Thomson eds. *Local Natures, Global Responsibilities: Ecocritical Perspectives on the New English Literatures*. New York: Rodopi, 243–255.

Boyle, D. (2002) *28 Days Later*. [Film] UK: Fox Searchlight Pictures.

Bridges, J. (1979) *The China Syndrome*. [Film] United States: Columbia Pictures.

Clark, T. (2015) *Ecocriticism on the Edge: The Anthropocene as a Threshold Concept*. London and Worldwide: Bloomsbury.

Cuarón, A. (2006) *Children of Men*. [Film] UK: Universal Pictures.

Davis, H., & Todd, Z. (2017) On the Importance of a Date, or, Decolonizing the Anthropocene. *ACME: An International E-Journal for Critical Geographies*, 16(4), 761–780.

Deleuze, G. & Guattari, F. (1987). *A Thousand Plateaus: Capitalism and Schizophrenia*. London: Bloomsbury Publishing.

Dixon, W. W. (2003) *Visions of the Apocalypse: Spectacles of Destruction in American Cinema*. London and New York: Wallflower Press.

Emmerich, R. (1996) *Independence Day*. [Film] United States: 20th Century Fox.

Emmerich, R. (2004) *The Day after Tomorrow*. [Film] United States: 20th Century Fox.

Emmerich, R. (2009) *2012*. [Film] Hollywood: Columbia Pictures.

Fiennes, S. (2006) *The Pervert's Guide to Cinema*. [Film] UK: Channel 4 Television Corporation.

Fox Broadcasting Company (2010) *The Walking Dead*. [Television]: Fox Broadcasting Company.

Graham, J. (2017). *Fast*. Manchester: Carcanet.

Hall, D. (2011) Varieties of Zombieism: Approaching Comparative Political Economy through *28 Days Later* and *Wild Zero*. *International Studies Perspectives*, 12(1), 1–17.

Hantke, S. (2010) Bush's America and the Return of Cold War Science Fiction: Alien Invasion in *Invasion*, *Threshold*, and *Surface Journal of Popular Film and Television*, 38(3), 143–151.

Joon-ho, B. (2013) *Snowpiercer*. [Film] South Korea: CJ Entertainment.

Kaplan, E. A. (2016) *Climate Trauma: Foreseeing the Future in Dystopian Film and Fiction*. New Brunswick, NJ and London: Rutgers University Press.

Klein, N. (2007) *The Shock Doctrine: The Rise of Disaster Capitalism*. London: Penguin Books.

Kopenawa, D. (2013) *The Falling Sky*. Cambridge, MA: Harvard University Press.

Kubrick, S. (1964) *Dr. Strangelove or: How I Learned to Stop Worrying and Love the Bomb*. [Film] UK: Columbia Pictures.

Lacan, J. (2006) *Écrits*. New York and London: W. W. Norton & Company.

Lawrence, F. (2007) *I am Legend*. [Film] United States: Warner Bros. Pictures.

Lebeau, V. (2001) *Psychoanalysis and Cinema*. London and New York: Wallflower.

Metz, C. (1982) *Psychoanalysis and Cinema: The Imaginary Signifier.* Basingstoke and London: Macmillan Press.
Mitchell, T. (2007) *Flood: A City on the Brink of Annihilation.* [Film] UK: Lionsgate.
Nolan, W. F. & Johnson, G. C. (1967) *Logan's Run.* New York: Bantam Book.
Petersen, W. (1995) *Outbreak.* [Film] United States: Warner Bros. Pictures.
Platts, T. K. (2013) Locating Zombies in the Sociology of Popular. Culture *Sociology Compass,* 7(7), 547–560.
Shortland, M. (1987) Screen Memories: Towards a History of Psychiatry and Psychoanalysis in the Movies. *The British Journal for the History of Science,* 20(4), 421–452.
Soderbergh, S. (2011) *Contagion.* [Film] United States: Warner Bros. Pictures.
Svoboda, M. (2016) Cli-Fi on the Screen(s): Patterns in the Representations of Climate Change in Fictional Films *WIREs. Climate Change*, 7, 43–64.
Swyngedouw, E. (2010) Apocalypse Forever? Post-political Populism and the Spectre of Climate Change. *Theory, Culture & Society*, 27(2–3), 213–232.
Swyngedouw, E. (2013) Apocalypse Now! Fear and Doomsday Pleasures. *Capitalism Nature Socialism*, 24(1), 9–18.
Troxler, H. & Pearsall, S. (1996) Audience Cheers When Aliens Destroy s House. *South Coast Today* [online], 5th July 1996. Available at: https://www.southcoasttoday.com/article/19960705/news/307059926
Turner, George. (2013) *The Sea and Summer.* London: Gollancz.
Whyte, K. P. (2018) Indigenous Science (Fiction) for the Anthropocene: Ancestral Dystopias and Fantasies of Climate Change Crises. *Environment & Planning E: Nature and Space,* 1(1–2), 224–242.
Williams, E. C. (2011) *Combined and Uneven Apocalypse.* Winchester and Washington: Zero Books.
Wright, E. (2004) *Shaun of the Dead.* [Film] UK: Universal Pictures.
Zalasiewicz, J., Williams, M., Haywood, A. & Ellis, M. (2011) The Anthropocene: A New Epoch of Geological Time? *Philosophical Transaction of the Royal Society A*, 369(1938), 835–841.
Žižek, S. (1999) *The Ticklish Subject: The Absent Centre of Political Ontology.* London: Verso.
Žižek, S. (2009) *First as Tragedy, Then as Farce.* London and New York: Verso.
Žižek, S. (2011) *Living in the End Times.* London and New York: Verso.
Zumbansen, N. & Fromme, M. (2010) Ecocatastrophies in Recent American (Non-)Fictional Texts and Films. *In:* L. Volkmann, N. Grimm, I. Detmers & K. Thomson eds. *Local Natures, Global Responsibilities: Ecocritical Perspectives on the New English Literatures.* New York: Rodopi, 273–287.

1 They say "our house is on fire" – on the climate emergency and (new) Earth politics

Edward H. Huijbens and Martin Gren

> DID YOU KNOW YOU HAVE A CALLING?
> An epic calling.
> A heroic calling.
> It's probably grander than anything you had let yourself imagine, outside of your dreams. You are *supposed* to save the world. That's why you are here, alive in this time of great consequence.
>
> – Margaret Klein Salamon (2020, p.xiii)

We live in troubled planetary climatic times. There is, unfortunately, no doubt about that. Abrupt climate change is now playing out on the earthly scene at a pace viscerally comprehensible to us humans and our fleeting existence. At the time of writing, May 2020 was the warmest month on record with the highest level of carbon dioxide in the atmosphere ever directly measured (417.16 ppm). The year is on track for becoming also the warmest in a series of record-breaking years since the start of detailed meteorological measurements. This heating of the planet is but one of numerous climate warning signs. We could recount shifting jet stream patterns with resulting changes in precipitation and heat distribution, wildfires raging over Arctic permafrost and the melting of the Greenland ice sheet at a rate that was previously not supposed to be seen for another few decades or so. Another worry is the ongoing deforestation in the Amazon that could move the rainforest towards a potential tipping point where it may irreversibly turn into a savannah, and according to climatologist Michael E. Mann it "is conceivable that much of Australia simply becomes too hot and dry for human habitation" (Mann, 2020a). The dire state of the planetary situation includes concerns over food production and biodiversity. Maria-Helena Semedo of the Food and Agriculture Organization (FAO) told a forum marking World Soil Day in 2014 that the world could run out of topsoil in about 60 years (see Arsenault, 2014). Insect populations are declining, and we are losing biodiversity at an unprecedented rate (Sánchez-Bayo and Wyckhuys, 2019). Even the prospect of an emerging sixth mass extinction is a scientific matter

DOI: 10.4324/9781003128854-2

of fact that we have as a matter of concern on our planetary table (Ceballos, Ehrlich and Dirzo, 2017; Kolbert, 2014).

Adopting the words of the climate activist Greta Thunberg, we can safely say that "our house is on fire" which is a way of recognizing the empirical scope and gravity of the ongoing planetary climate mutation. If we are to avoid catastrophic climate change and its devastating consequences for humanity, we need to immediately take climate action that is planetary in scope. In order to stay below a 2°C increase in global average temperatures compared to pre-industrial levels, as stipulated by the Paris Accord, we have very limited time at our disposal. Although many figures are floating around of the fabled "window of opportunity" that is supposedly still open to us, they tend to gravitate around a decade or so. What is required to happen during this short period of time is then nothing but a historically unprecedented revolutionary overhaul of the very fabric of our societies, and at planetary scale. As Greta Thunberg puts it: "I want you to act as if our house is on fire. Because it is" (Thunberg, 2019, p.24).

To say that "our house is on fire" is also to recognize that the Earth and humans are critically trans-mutating through a common planetary "super wicked" problem, whereby time is running out with no planetary authority to address the problem, coherent will to change and a future discounted in policy suggestions (Levin, et al. 2012). Speaking as geographers, we note that in modern human geography the Earth has primarily been conceptualized as Earth's surface on which humans as social subjects have made their spatial imprints. We will argue that this modern socio-spatial theorizing, amplified by a political cultural agenda, is insufficient and obsolete when we now are facing the wicked Earth of the Anthropocene. On that planetary level we now need to conceptualize humans not only as geographical subjects but also as a geo-force affecting the functioning of the Earth System. However, both humans and the Earth also turn out to be problematic at that scale. Accordingly, we want to scrutinize conceptualizations and theorizations of the Earth and humans that are more adequate and relevant to our present earthly planetary state of climate emergency and its urgent call for climate action. Here we focus on humans as earthly beings, dwelling in a house on fire in the midst of an emerging Apocalypse in search for its earthbound politics. How can we come down to Earth and land with a growing population and demands for resources in the midst of the climate emergency? Can the Apocalypse heed the calling to save the world?

We begin this chapter with how the climate emergency equates to a house on fire. In the next section, we therefore present some current empirically grounded insights as to what is burning and the emergency of the climate situation, including calls to urgent action. Then there is the question about the nature of the house that is on fire, which here implies that the Earth can be conceptualized in many different ways, for example as "humanity's common planetary house", the "Earth's surface" or the "Earth System". These are big Earth concepts that are useful at the planetary scale, but they can also be paralyzing for Earth politics and concrete climate action. We therefore break these down in order to illustrate how we can give figure to the Earth as "Critical Zone" (CZ), and with some occasional references to "the Terrestrial" as earthly demarcation human habitation. With this focus we propose ways in which we can compose our planetary house in actionable terms, carving out a politics of earthly habitation that does not *a priori* distinguish between nature and society, nor separate the old social question from the earthly local and planetary climate and ecological conditions. The emerging Earth politics in the fourth section of the chapter thus highlights the "planetary vital signs" we want to see incorporated into an earthly climate politics of the CZ. We end the chapter with some thoughts on paradoxical life in the earthly house of the looming Apocalypse – our dwelling in troubled earthly planetary end times.

Climate emergency: when our house is on fire

In the year 2019 "climate emergency" made it to the headlines and became established in various fields within and outside academia. A network of 7,000 universities declared a climate emergency (O'Malley, 2019), and so did the European parliament while it also urged all EU countries to commit to net zero greenhouse gas emissions by 2050 (Rankin, 2019). The climate activist Greta Thunberg became Time's person of the year, and Extinction Rebellion (XR) established itself as a major climate activist group. XR's aim is to rebel against extinction, not just of humans and polar bears but of all earthly things and creatures, such as glaciers, insects and forests. This year (2020) we have witnessed more than 11,000 world scientists declaring a "climate emergency" (Ripple et al., 2020).

While people rebel and academics, individuals and supra-national assemblies declare a climate emergency, the time for action is now. Climate emergency is a pressing imperative to act while there is still a chance to turn the tide of climate and ecological crises sweeping the planet. Most importantly, climate action has to be conducted on such a grand scale that small individual contributions, however well intended, are doomed to be insufficient. In Greta Thunberg's words: "Everything needs to change. And it has to start today" (Thunberg, 2019, p.12). According to the UN (2019) report *United in Science*, "policies to lower emissions must *triple* to meet the 2 degrees Celsius limit, and *fivefold* to align with the 1.5 Celsius limit" (p.16 *our emphasis*). The time for climate mobilization is now, not because of all the talk about climate emergency, but for another reason: *we are living in it.* Our house is on fire, and every single day of (in)action counts.

Our planetary house heats up from the inside, and the principal climate mitigation strategy is to rapidly reduce the burning of fossil fuels, typically with the goal of halving our emissions by 2030 against 1990 levels. When the European Parliament announced its recognition of the climate emergency, these reduction targets were adjusted for the EU from 40% to 55%. Yet, as

Anderson, Broderick and Stoddard (2020) suggest, just the decarbonization targets of Britain and Sweden would produce emissions between two and three times the carbon budget required to meet the goals stipulated in the Paris Accord (i.e. limit global warming to 2.0°C and preferably 1.5°C).

> Even within more 'climate progressive' nations [like Sweden and Britain], the Paris Agreement necessitates an immediate increase in their proposed mitigation rates by a factor of two to over 10% p.a., with full decarbonization achieved across all sectors by 2035–40. Delivering such rapid and deep mitigation implies profound changes to many facets of contemporary industrial society. But failing to take appropriate action will increasingly lock-in devastating climate impacts, imposed initially on poor and climate vulnerable societies, but ultimately across all of the international community and natural ecosystems.
>
> (Andersson, Broderick and Stoddart, 2020, p.12)

In the context of emergency, then, full decarbonization needs to be achieved by 2035–2040, and not even the targets set by climate progressive nations are sufficient. Of note is that targets investigated refer to plans that often tend to be optimistic, and where there is a long way to implementation. When it comes to declaring climate emergency, the situation may not be that different. A declaration in words, however well intended, will not reduce the amount of carbon dioxide in the atmosphere; neither will a signature on a Paris Accord stop the heating of the planet. In fact, under business-as-usual scenario we seem to be on track for at least 3°C of global warming, and that is likely to be a conservative estimate by most accounts. The consequences at 2°C are expected to be:

> that more than 150 million additional people would die from the effects of pollution, storms that used to arrive once every century would hit every single year, and that lands that are today home to 1.5 billion people would become literally uninhabitable, at least by the standards of human history.
>
> (Wallace-Wells, 2020, see also: Xu et al., 2020)

In a widely acknowledged paper from 2018, "Trajectories of the Earth System in the Anthropocene", the authors suggest that 2°C may in fact be a potential threshold that could activate dangerous tipping points that could take the Earth System towards even higher temperatures. "[E]ven if the Paris Accord target of a 1.5C to 2.0C rise in temperature is met, we cannot exclude the risk that a cascade of feedbacks could push the Earth System irreversibly onto a 'Hothouse Earth pathway'" (Steffen et al. 2018, p.8254). In that perspective, the Paris Accord goal of keeping global warming below 2°C looks more and more like a planetary mission impossible. If that were to be true, then the Earth politics of the Paris Accord actually appears to be an issued death sentence for the human species rather than a document for securing humanity's common planetary future. In anticipation of the forthcoming IPCC report of the global climate due out 2021, one of the lead authors Dr Joëlle Gergis cannot but wonder:

> …whether the Earth system has now breached a tipping point, an irreversible shift in the stability of the planetary system. There may now be so much heat trapped in the system that we may have already triggered a domino effect that could unleash a cascade of abrupt changes that will continue to play out in the years and decades to come. Rapid climate change has the potential to reconfigure life on the planet as we know it.
>
> (Gergis, 2020)

Declaring climate emergency implies an unprecedented planetary climate action mobilization and transformation, often articulated as avoiding a global heating beyond 1.5–2°C. However, according to Mann, "1.5C might be impossible now (without artificial sequestration technology). We go for the earliest exit ramp we can" (Mann, 2020b). Climate emergency also suggests something completely different from climate mitigation and adaption steeped in the gradualist modern discourse of sustainable development. Emergency is literally a matter of life and death, looking for the earliest exit ramp and saving as much as possible while the house is on fire. It is perfectly understandable that so many of us often shy away from the existential planetary threats that we, as humanity or as species, are facing. Human, all too human. As the climate and ecological emergency unfolds an internal battle is raging, as if we are torn between reasoning and emotioning. As Salamon, who pioneered the climate emergency declaration campaign, puts it:

> We sense we're in climate emergency and mass extinction event, but we have a deep-seated psychological instinct to defend against that knowledge. The pain is shouting at us: "Everything is dying!" Somewhere inside, we feel the horrors of civilizational collapse and the sixth mass extinction of our species, in our bodies. Our minds attempt to shield us from this pain – we avoid, distract, deny, and numb ourselves. But these defences work only temporarily: When we fail to process our emotions and mourn our losses, the pain takes on tremendous power. It follows us around like a shadow, and we become increasingly desperate to avoid what we know.
>
> (Salamon, 2020, p.2)

We have no intention here of going into Freud, but living under the murky spell of climate and ecological emergency evokes all kinds of existential, emotional and cognitive dissonances and reactions. On that psychological canvas floats the temporality of emergency in the present. A climate revolution has to take place at planetary scale within ten years or so, the precise numbers are here not that important. As Mann (2020a) puts it: "We are the blindfolded man who is told he is nearing the edge of a cliff. Is he three steps away? Four? Ten? Regardless of the distance, his only safe course of action is to stop lurching forward". Regardless of whether "we" is a man, woman, he, she, humanity or a human species, to stop lurching forward requires, according to the climate emergency advocates, a massive collective climate action and unprecedented radical transformations of the very fabric of our societies. If we fail, those among us who have already been punished by history will be among the first to be punished yet again.

Declaring climate emergency, that is, to speak and act on the premise that our house is on fire, raises several questions about what it is that is on fire. The emergency refers to the state of climate relative the planet, and that will inevitably take us to the Earth. However, the Earth can be conceptualized in many different ways, for example as "earth surface" and "Earth System". Many more possible articulations of the Earth exists. such as David Abram's framing of all our encounters as "telluric", whereby the Earth is expressed through enacting itself by "expressive magic in its own manner… [as] a property of animate earth itself" (Abram, 2010, p.171). The language of poetry and the voices of novelists and storytellers are mediations of modes of existence that can be from the past, present or the future. These are abundantly available to jolt our sensibilities in new directions and open us to different possibilities to give figure to the Earth at large, including our own earthly agencies. In every culture and every place there are possibly ideas around of a different relationship with our surroundings. Déborah Danowski and Eduardo de Castro (2016) explain for instance with fervour how indigenous Amerindian creation myths allow for earthly practices without endless progress and have prevailed in the face of the devastation of their populations and cultures. All these ways of conceptualizing the Earth and our relations with "it" have consequences for our climate thinking and actions. "Saving the planet" is not the same as conserving soil and insects on a small tract of land. What is the Earth and what is "we"? How does alternative earthly sensibilities and practices translate to loss of biodiversity and sea-level rise? Can our common murky and malleable planetary waters be charted in ways that are at the same time also local and Earthbound?

The Earth: what house is on fire?

We humans have made such fundamental changes to our planet that one can make the argument that it also needs a new proper name. Among the possible ones we find Bill McKibben's "Eaarth" (McKibben, 2010). Perhaps for understandable reasons it did not catch on, and now the Earth is increasingly understood in the context of the Anthropocene (i.e. the proposed new geological epoch distinguished by the geo-force of humans, *the Anthropos*). The Anthropocene comes with a particular new kind of Earth, one that has moved from background to foreground. This Earth simultaneously alters our understanding of Society and what it means to be human.

We might here recall Galileo who, as part of the Copernican revolution, claimed that "the Earth moves". Today, four centuries later, in the Anthropocene, the Earth indeed moves again, but in a completely different way. In addition to its celestial motion, a planet orbiting around the sun, we now understand that the actions of humans "move" the Earth, and in ways that are faster and more widespread than we could ever imagine. The Earth is responding to our actions in real time. When Galileo and the astronomers made the Earth move around the sun, "the whole fabric of society [also] felt under attack" (Latour, 2020a, p.1). Even though the motion of the planet could not be perceived in everyday life, it nevertheless had real impact on the religious and social order. Now the moving Earth of the Anthropocene calls for a (new) Earth politics.

When geologists and geoscientists of the Anthropocene now transform the Earth into a moving *Earth System*, the whole organization of society, and what it is made of, is also being subverted. By definition the Anthropocene states that the trajectory of this geological epoch cannot be separated from the activities of humans (the Anthropos). With this understanding we can detect that humans also move the Earth System: melting glaciers, jet streams that meander, wildfires in the Arctic, city-sized swarms of locusts that are wreaking havoc and what else of earthly-planetary "movements" there may be. In fact, the Earth System is actually mutating so fast that it now overlaps with our human time-scale, and thereby effectively braiding our respective agencies. Indeed, we are confronted with an Earth System that is no longer changing only in the slow pace of its own geological time. It now moves at a pace even faster than human history, and we too have to act fast if we want the Earth System to remain in a "safe operating space for humanity" (Rockström et al., 2009). Climate emergency means in the Anthropocene that "[u]rgent action is required to avoid further collapse of the Earth System" (Morris, 2019, p.55).

However, the Earth of the Anthropocene, and particularly in relation to the climate and ecological emergency, has thus far been rather poorly conceptualized in the social sciences and the humanities. In modern thinking the Earth has most often been regarded as a passive backdrop for social life, a stage for the big actors of Nature/the environment and Society/culture.

> Throughout recent history, an underlying stable condition of the Earth System has been taken as a given. This is the premise upon which our legal and political structures have been created over the past several centuries. /…/ there has been an implicit assumption that current conditions form an objective and unchanging reality that has surrounded us since time immemorial.
>
> (Zalasiewicz et al., 2019, p.36)

This modern understanding corresponds to the Earth in the geological epoch Holocene (the last 12,000 years or so), where it was a relatively stable dormant agency which was not supposed to wake up and change the planetary course before the next ice age. In modern thought the Holocene Earth has been re-presented as the Earth's surface, which has also served as a primary reference plane for politics. This can be illustrated by modern geo-politics where states were considered to be territorial units. Conflicts and wars were basically about their demand and control over space on the Earth's surface, and did not involve the kind of Earth that we have come to know and be dependent on in the Anthropocene and in the climate emergency. For example, in the so-called "world wars" the "Earth was the board on which conflicts were waged, not a party of those conflicts" (Latour, 2020a, p.5). We can here note that in modern thinking another abstract Earth has also been over-layering the Earth's surface, namely "the Globe".

> …there has been confusion between the Globe and the Earth. Such a Globe is still the undisputed, authoritative, universal, external frame inside which all geopolitical entities – be they empires, nation-states, lobbies, networks, international organisations, corporations, diasporas – are situated in a recognisable place, a province side by side with all the other provinces. In other words, a natural Globe still offers the 'ground map' which allows any localisation to occur.
>
> (Latour, 2016, p.307)

As we move from the Earth of the Holocene to the Earth of the Anthropocene, a conflict between the Globe and the Planetary also arises. There have been many wars in human history, but none has engaged the planetary *as such* (Chakrabarty, 2009; Latour, 2020a).

> The First World War had generated a novel idea of the global horizon, but entirely failed to let the "Planetary" emerge as such: yes it was a world war, but the planet was still taken as a single checkerboard for human players.
>
> (Latour and Chakrabarty, 2020, p.7)

Now that the stable conditions of Holocene Earth has disappeared, and with the Earth System in the driving seat in the Anthropocene, we are in the hands of an unruly defiant Earth (Hamilton, 2017). During the Holocene humans were, in principle, only a *geographical* force. Their activities transformed the Earth's surface, environments, places, landscapes, nature, but not the Earth System itself, and the Earth System most certainly did not speak back in any way we as humans in our fleeting existence could grasp or comprehend (Zalasiewicz et al., 2019). The message of the Anthropocene is that humans are also a collective *geological force* that affects the functioning of the Earth System. The melting of the glaciers is partly caused by human action, and their melting will in turn inevitably have consequences for our own possibilities for earthly habitation. Indeed, geography's rationale as "earth-writing" seems now to have gone full circle, as the intruding agency of a mutating planetary Earth is now also literally writing us (Gren, 2017). At present we are nervously waiting for an all too soon blue ocean event up in the Arctic, and the dire planetary consequences thereof.

Yet, humans remain also in the planetary driver's seat, albeit with a fundamental difference. We can no longer operate on the modern assumption that we can clearly separate our own agency and historical temporality from the geological temporality and agency of the Earth System. Humans and the Earth System are instead two intertwined parts of a common "geo-story", and they co-author each other's destinies in real time (Chakrabarty, 2009). Gone also are the modern days when freedom was about human relationships in social isolation, because "as we enter the Anthropocene freedom must also, and primarily, be understood as it bears on our relationship with the Earth" (Hamilton, 2017, p.150). As the motto of Sloterdijk's anthropology reads, "*Tell me what you are immersed in, and I will tell you what you are* [emphasis in original]" (Sloterdijk, 2016, p.17).

As for us humans, it may be *heimlich* to be part of an Anthropos and feel a familiarity at the species level. Yet, everybody knows that the Anthropos does not correspond to a political subject or agency that by a magic earthly wand can turn down the heating of the planet. And controlling the so-called "Earth System" is as eerie as subverting the so-called "capitalist system". We might say that the Anthropocene is a calling to come back down to Earth, but we also need to recognize that the big co-ordinates of Earth System and the Anthropos will only get us so far. To which Earth are we supposed to get back? On which Earth should we land (Latour, 2018a)? Who are the "we"? That we live on the Earth may seem obvious and self-evident. Yet, one could in fact also claim that we do not live *on* the Earth, but that we instead live *in* something else.

The CZ was defined in 2001, but has only recently been acknowledged as a distinct co-evolving entity driven by physical, chemical and biological processes that sustain life. The CZ includes atmosphere, water, biology, regolith, land surface, and is recognized as an entity composed of co-evolving systems that create the structured dynamic skin of the Earth (Brantley et al., 2017, pp.852, 856). A few kilometres down and a few up relative to the Earth's surface at median sea level is where all terrestrial life exists. It is in this thin varnish, the skin of the Earth, where life produces and maintains itself. At present we do not have a good conceptual vocabulary for describing the Earth of the CZ, not even in geography. One of few geographers who have systematically tried to develop a conceptual apparatus for mapping the conditions of the CZ, although he did not call it that, was the Swedish geographer Torsten Hägerstrand (well known for his time-geography). In his final book, unfortunately only in Swedish, he develops a geography centred on "tillvaroväven" ("the web of becoming"), or "the fabric of geographical co-existence" (Hägerstrand, 2009, see also 1976). The web of geographical becomings is composed by all earthly creatures and their co-evolving geographies, thereby highlighting the time-space signature of what Moore (2015) and Glacken (1967) would call the "web of life".

Our aim here is not to provide a detailed account of the CZ, and there are of course other ways of downscaling the Earth, most notably in this context are "the biosphere" and "the geobiosphere". Whatever the Earth, our climatic vulnerabilities need to be recognized, politicized and negotiated (Hamilton, 2015). The point is that CZ offers one alternative to big concepts like Earth System, the Anthropocene and Humanity. As necessary as it may be to try to "save the planet", to keep the "Earth System" in a "safe operating space for humanity" or to focus on the "planetary scale", we are still operating on a level for Earth politics that is too remote from earthly human practice. It is arguably in something like the CZ (or "web of becoming", or "the fabric of geographical co-existence") that we as a geographically differentiated terrestrial being will have to land on, as it denotes the common planetary house of life we inhabit. The climate of this house (the skin of the Earth in which we reside) is simultaneously co-constituted by our geographically specific territories and territorialities.

It follows that the CZ becomes an earthly entity that can and needs to be politically and spatially re-composed, especially so in times of planetary climate and ecological emergency. Indeed, if ever there was a time for earth writing and speaking an earthly language, this is the one. The "earthly imperative" in the Anthropocene and in the climate emergency is to pave the runway of where to land. This requires that fragile possibilities are underpinned by an earthly, or terrestrial, politics that can help us navigate our present state of planetary unsustainability. It needs to revolve around a political mobilization of the CZ in the here and now, of "the Terrestrial", or in other ways be able to align the hackneyed social question with the earthly.

(New) Earth politics: composing our planetary house

In the perspective of the Anthropocene, any political composing of our planetary common house has to take the braided collective geo-force of humans and the functioning of the Earth System into consideration. As noted, this is easier said than done. The *Anthropos* of the Anthropocene, we humans, remains an abstract being, and there is no corresponding political assembly available where this being can settle its political climate disputes and compose its planetary commons. Likewise, the Earth System is also a too big of a concept, and it also comes with a particular natural science narrative of the Earth, even though that can give us some tools for identifying important planetary "vital signs" like for example "temperature, precipitation, river flow, glacier behavior, groundwater reserves, sea level, seismic activity" (Bjornerud, 2018, p.63).

Although "[t]he dramatic narratives of the geologic past are perfectly suited to the human appetite for storytelling" (Bjornerud, 2018, p.174), we need to hone in and break down units for political analysis and action, yet with an eye focused on the overall planetary climate and ecological emergency. In order to do climate Earth politics, we need some kind of earthly grounding that defines where we reside and where the web of life is sustained. For that reason, it is to something like the CZ we must go, as it denotes the thin skin of the Earth where all life is composed. The CZ, possibly further demarcated as the Terrestrial (Latour 2018a), becomes an earthly entity for climate politics as well as for climate action. This is at least something different from de-politicizing gesture of negotiating the Anthropocene from the stand-point of some kind of naturalized emergency, which;

> ...resides precisely in letting the naming of a geo-social epoch and a contingent "truth" of nature decide our politics, thereby disavowing that the "our" or "the human" does not exist.
>
> (Swyngedouw, 2019, p.256)

Furthermore, any attempt to do politics in and of the CZ, or what Latour is referring to as "terrestrial politics" (Latour, 2018b, 2020b), must grapple with how to combine the old "social question" with an understanding of humans as earthly beings composing life together with all other beings in the CZ. It is of little help, especially in the light of planetary climate and ecological emergency, if our politics lead to the improvement of social justice while, for instance, the current rate of soil depletion continues. In the same vein, getting a political climate grip on the soil and the land, where we are born and through which we are allowed to continue to exist, will make little progress if we do not simultaneously address social and economic issues of control and ownership.

One could argue that modern political thought has been biased towards the social, and for understandable reasons. When Karl Marx developed his theory of historical materialism by the end of the nineteenth century, he could see with his own eyes that the production of material wealth was tied to awful social and economic injustices. In volume one of *Capital* he stated that labour was the "father" of material wealth, but he also noted in passing that it had "earth as its mother" (Marx, 1887, p.31). Nevertheless, the Earth remained in the background for Marx, as an Earth surface offering a geographical distribution of material resources. Marx came to focus on the politics of the social world that we live *in*, and on the social class struggle that comes with it. "If the accumulation of capital is the proletarianization of labour, it is also the production of knowledges aimed at controlling, mapping, and quantifying the worlds of commodification and appropriation" (Moore, 2015, p.20, citing Marx's *Capital* on labour).

As Jason Moore however highlights: "At the core of this law is the ongoing, radically expansive, and the relentlessly innovative quest to turn the work/energy of the biosphere into capital (value-in-motion)" (p.14). In other words, the "...work/energy of the web of life is incorporated into the relations of power and re/production" (Moore, 2015, p.15). In the Anthropocene, we are facing another class struggle that is not only social since it now also involves the geological.

> Geology is real enough. But it becomes *geo-history* through definite relations of power and production in which geological dispositions are immanent. Geology cannot "directly determine" the organization of production, precisely because production relations are co-produced.
>
> (Moore, 2015, p.44, emphasis original)

The relationship between the social and the geological in the Anthropocene gives rise to what one may call a "geo-class struggle". We are dealing with "geo-social formations", which are saturated with both anthropological and geographical differences (Clark and Yusoff, 2017). We can think of the fact that as some of us opt not to fly and even more cannot afford it, Bill Gates may be up in the air for weeks during a year. This is not only a socio-economic class issue, him being superrich, but also a question about our respective ecological footprints and their earthly consequences. This becomes a "geo-class" struggle which is less about the unequal access to wealth *in* society and more about the Earth Bill Gates and we all live *off*. This is also reflective of an earthly rift between the particular territory a state or country occupies on the Earth's surface, and the territorialities it and its citizens depend on for their subsistence. As Latour points out, there "is a world *in which one lives*, the one that has justice, rights and obligations, the vote, citizenhood; and there is a world *one lives off*, which has become a very way off, down below" (Latour, 2019, p.9). In the "new climatic regime", which Latour uses in order to give legal and institutional dimensions to the Anthropocene, we are all torn between the world we live in and the world we live off.

In terms of both climate and ecological emergency it becomes evident that the earthly world that we live *off*, or *from*, no longer provides us with a stable immutable ground for accumulating our material wealth, nor for engendering our ecological habitats. One could say that this "*earthly* earth-world" we live *from* is now irrupting in the midst of the "*social* earth-world" we live *in*. The overarching geo-political challenge in the Anthropocene then becomes how these two earth-worlds, and these two class struggles, can be reconciled *and* how they could mobilize a climate politics and climate action that can help us to find our way, not so much *on* planet Earth, but *inside* the CZ, and therein the enacted demarcation of the Earth as the planetary common ecological zone that is critically relevant for human survival. Underneath the Earth System are insects and soil, underneath the capitalists are workers, and "underneath the workers are living things!" (Latour, 2020b, p.8). This points to the important observation that the ecological must be added to the class struggle between social and geological.

A terrestrial politics *in*, *of*, *for*, the CZ, faces daunting difficulties. One of them is how to combine social justice with climate justice, to which we would add ecological justice. In the words of Pope Francis, we need to politically detect "*both the cry of the earth and the cry of the poor*" (LS, p.49, cited in Latour, 2018b, p.5). The CZ, as well as the planetary climate and ecological emergency, does not sit very well inside the political space of the nation-state. For example, "[w]hen you say it's yours, do you include the red sand blowing from Sahara or the acid rain from Chinese factories?" (Latour, 2020, p.9). In terms of scale, the CZ, for human habitation, is always terrestrially local, but every pocket of local order is at the same time co-dependent on territorialities that take it all the way to our planetary common CZ. We can try to take care and protect our own trees, but what about the forests that belong to other territories and other people? However, geographically distant, their territorial politics are also part of the composition of other terrestrial territories. If Bolsonaro decides to instigate actions that turn the Amazon rainforest into savanna, and Trump pulls out of the Paris Accord, are they not actually declaring a geo-political climate war on *us here*? Are they not effectively saying that "We don't want to live on *your* planet!"? Transformed into action, they undermine the prospects for us to live on *our* planet. At the time of proofreading we now know that US has re-joined the Paris Agreement, perhaps that could be read as a gesture of peace in "the new climate war" (Mann 2021)?

So it is that our common planetary home is fraught with earthly political tensions, and some of its occupants have even locked themselves up in their own rooms imagining that a common habitable Earth has nothing to do with them and think they can fireproof their own walls as the rest burns to the ground. As Bjornerud states, "…our current society is a kleptocracy stealing from the future" (2018, p.165). Latour spares no punches in telling us that the loosely defined elites, carrying the emblem of Donald Trump, have already;

> given up the idea of actually pursuing the modernization of the planet with everyone, because they knew, before everyone else, that such modernization was impossible – precisely for want of a planet vast enough for their dreams of growth for all.
>
> (Latour, 2018a, pp.22–23)

We all live in certain rooms in the planetary house, subjected to its nooks and crannies of which the nation-state is a terrible signifier. With the dawning realization of planetary boundaries being crossed, the prospect of a dire planetary future is now thrown at us all, playing a ghastly role in our attempts to figure out who we are as political actants and how we can politically proceed with dignity in terms of an ecologically expanded understanding of "geo-social justice" (Clark and Yusoff, 2017). Fire proofing the planetary house is now a task for all of us, but it will mean different things to the different rooms we happen to inhabit, and their crannies to our common home. Yet fireproofing is not the same as conducting a rescue operation when a house is on fire. Doing Earth politics under emergency conditions means to as quick as possible save as much as possible. As the COVID-19 made abundantly clear, whichever way the Earth responds to our activities, in this case a zoonotic disease, the ramifications play out in our societies and through the extent of our reactions. In this extraordinary case, what everyone before thought to be impossible, the economy and its growth engine actually had an emergency off-button.

> A hiatus of just two months is all it took to achieve what numerous studies by sociologists of markets and anthropologists of finance would never have achieved: a widely-shared realization that the economy holds in place only as long as the institution that performs it – and not a day longer.
>
> (Latour, 2020b, p.5)

Although still in the throes of the COVID-19 pandemic, and the case is still open as to how it will play out, what to us is clear is that, in spite of the climate and ecological emergency, the political pressure now is to push the on-button and go back to "normal" as quick as possible. The politics of what is currently understood as "the economy", which means that its ecological roots have been amputated, thus casts its dark purgatory shadow over the Earth as the place for human habitation.

Inside the house of the Apocalypse: dwelling in troubled earthly end times

So how to go ahead, while dwelling inside a house on fire? How to find one's way when action on "fire", as climate and ecological emergency, is still being stalled by prolonged investigations inside various imaginary departments of fire and emergency management, all with addresses quite unknown? Moreover, what to do in times when there is not even a common understanding of "the house" in question? In Latour's words, "[t]he great tragedy of the present situation is that there is no agreed upon definition of which planet we are supposed to inhabit in common" (Latour, 2020, p.13). Perhaps this also suggests that if one wants to understand human thought-and-action, then there is much to learn from Greek tragedy (Olsson, 2007). Everything looks fine in the beginning, all the actors (read: "the Moderns") are filled up with good intentions (read: "the promises of Modernity"). However, as they make progress along their way to a future Utopia (read: "an imaginary place without real earthly grounding"), and thereby manage to emancipate themselves from their earthly attachments, they eventually end up in a terrible terrifying planetary bondage (read: "they managed to free themselves from Nature but are now waking up enslaved by the Earth System"). All in all, the accumulation of unintended consequences at planetary scale. By taking action in accordance with their own maps, with the co-ordinates of

Society and Nature writ large, the Moderns were supposed to gradually move towards a better "Future". Now, in the Anthropocene, in the climate and ecological emergency, they instead find themselves suspended in mid-air, realizing that they will have to come back down to Earth. Meanwhile, the speed of a dire planetary climate and ecological future moving towards them correlates all too well with the temporality of their own climate (in) actions.

Given the scientific evidence of climate and ecological change at planetary scale, it is impossible to arrive at any other conclusion than that "our house is on fire". Consequently, there is no doubt that a historically unprecedented transformational climate and societal change is urgently required in order to avoid at least some of the devastating consequences for humanity at planetary scale. We live in climate and ecological emergency (not to mention all other possible emergencies), which means that humanity, as we have come to know it, is facing a literal existential threat, to our civilization and even to our species (i.e. sixth mass extinction event). No wonder that references to "the Apocalypse" now seem to abound. Apocalypse has a somewhat complex etymology and is apparently used in different ways, but most often it seems to denote something more or less purely negative and disastrous. In that sense, it also comes with a connotation of "end times", but what about "the end" of what, and what about "the time"? One clue is suggested by etymology since the Greek word "apokalyptein" means "disclosure" and "revelation". In the Book of Revelation in the New Testament (also known as "Apocalypse"), the word "revelation" denotes God's will and a divine providence in relation to the end of the world and final judgement. The most fundamental appeal of religious apocalypticism seems to be the conviction it holds forth that time is related to eternity. The history of the human species/humanity has discernible structure and meaning in relation to its End, and this End is the product not of chance but of a divine plan. It is that plan (the will of God), and how humans have responded to it in their earthly practices, that will be the frame for judging them in the End time of the Apocalypse. However, one does not have to restrict "Apocalypse" to religious interpretations. In the context of the Anthropocene, and the climate and ecological emergency, the references shift from a divine providence, a heaven above and its temporal horizon designed by a God above, to existence on an Earth down below. The final judgement, for earthly human purposes, is then not made by God, but by the Earth (however conceived). The end time formula, which is central to religious apocalypticism, can then be transformed into a secular variant. In relation to climate and ecological emergency, our earthly present planetary situation, we believe that two positions can be identified for heuristic purposes. In the *first*, the Apocalypse comes to denote that we do live in earthly end times, and that there is basically nothing we humans can do about it. The planetary climate system, in this case "the Earth" instead of "God", is too powerful, and has set us on a dire and inevitably disastrous trajectory. In the *second* position, however, it becomes instead important to accept that we actually live in the Apocalypse, precisely in earthly End times (see Malm, 2020; Williams, 2011). Now is the time to define ourselves, now is the time for taking actions that are worthy of an earthly judgement. This "earthly Apocalypse" also changes the notion of the future from divine to earthly, and reverses its direction to an earthly future that is also coming towards us. This plays out particularly against the Modern understanding where climate action (living and acting in the present) always can be postponed into an undefined utopian future that comes with the promises of delivering technological fixes (Hamilton, 2015). As long as we cling to the Modern hope, in the sense that some solutions in the future will save us (compare with "Gods will" and divine power over human destiny), we will not accept, nor take responsibility, for living in the time of the earthly Apocalypse, Perhaps this is somewhat in line with what Greta Thunberg is advocating:

> Adults keep saying: 'We owe it to the young people to give them hope'. But I don't want your hope. I don't want you to be hopeful. I want you to panic. I want you to feel the fear that I feel every day. And then I want you to act. I want you to act as you would in a crisis. I want you to act as if our house is one fire. Because it is.
>
> (Thunberg, 2019, p.24)

In our reading, it is our current planetary (dis)juncture that turns the original theological myth of the Apocalypse into an earthbound planetary climate and ecological crises. It is then not the final judgement of God, but the present earthly climate and ecological planetary order that reshapes the lives of human beings and where particularly the Earth-world as we know it comes to an End as modern technology continues its earthly destruction. However, the original meaning of Apocalypse also includes "disclosure" and "revelation". Disclosing the veil of the Earth-world can imply the demise of an older order and the creation of a new, whether this new is utopian or dystopian is another question. A new Earth politics will have to emerge to us from the earthly entanglements we are embedded in, which is something else than the magmatic rumblings of the Earth to the sun it encircles once a year, or by an Earth politics centred on Society and Nature. Faced with the earthly Apocalypse it is our primary task to listen and comprehend these expressions and allow them a role in making for our territories and territorialities. In the earthly Apocalypse there is little time for listening to the clamour of idealized progress and hope for grand fixes to the Earth System by an abstracted humanity, or inversely the doom and gloom of a religious End time. All our political declarations, accords or lofty promises will not reduce the amount of carbon dioxide in the atmosphere or stop the heating of the planet. After all, especially in the earthly Apocalypse, *you* are indeed *supposed* to save the world now (tomorrow may already be too late).

References

Abram, D. (2010). *Becoming Animal: An Earthly Cosmology*. New York: Vintage Books.

Anderson, K., Broderick, J.F. and Stoddard, I. (2020). A Factor of Two: How the Mitigation Plans of 'Climate Progressive' Nations Fall Far Short of Paris-Compliant Pathways. *Climate Policy*, DOI: 10.1080/14693062.2020.1728209.

Arsenault, C. (2014). Only 60 Years of Farming Left If Soil Degradation Continues. *Scientific American*. From: www.scientificamerican.com/article/only-60-years-of-farming-left-if-soil-degradation-continues/, viewed 30 June 2020.

Bjornerud, M. (2018). *Timefulness: How Thinking Like a Geologist Can Help Save the World*. Princeton: Princeton University Press.

Brantley, S.L. et al (2017). Designing a Network of Critical Zone Observatories to Explore the Living Skin of the Terrestrial Earth. *Earth Surface Dynamics*, 5, 841–860.

Ceballos, G. Ehrlich, P.R. and Dirzo, R. (2017). Biological Annihilation via the Ongoing Sixth Mass Extinction Signaled by Vertebrate Population Losses and Declines. *PNAS Proceedings of the National Academy of Sciences of the Unites States of America*, 114(30), E6089–E6096.

Chakrabarty, D. (2009). The Climate of History: Four Theses. *Critical Inquiry*, 35(2), 197–222.

Clark, N. and Yusoff, K. (2017). Geosocial Formations and the Anthropocene. *Theory, Culture & Society*, 34(2–3), 3–23.

Danowski, D. and de Castro, E.V. (2016). *The Ends of the World*. Cambridge: Polity.

Gergis, J. (2020). We Are Seeing the Very Worst of Our Scientific Predictions Come to Pass in These Bushfires. *The Guardian*. From: www.theguardian.com/commentisfree/2020/jan/03/we-are-seeing-the-very-worst-of-our-scientific-predictions-come-to-pass-in-these-bushfires, viewed 30 June 2020.

Glacken, C.J. (1967). *Traces on the Rhodian Shore. Nature and Culture in Western Thought from Ancient Times to the End of the Eighteenth Century*. Berkeley: University of California Press.

Gren, M. (2017). Mayday – A Letter from the Earth. In: *The Question of Space: Interrogating the Spatial Turn between Disciplines* (edited by Marijn Nieuwenhuis and David Crouch). Lanham: Rowman & Littlefield.

Hägerstrand, T. (1976). Geography and the Study of Interaction between Nature and Society. *Geoforum*, 7, 329–334.

Hägerstrand, T. (2009). *Tillvaroväven*. Stockholm: Forskningsrådet Formas.

Hamilton, C. (2015). *Requiem for a Species. Why We Resist the Truth about Climate Change*. London: Routledge.

Hamilton, C. (2017). *Defiant Earth – The Fate of Humans in the Anthropocene*. Polity Press.

Kolbert, E. (2014). *The Sixth Mass Extinction. An Unnatural History*. New York: Picador.

Latour, B. (2016). Onus Orbis Terrarum: About a Possible Shift in the Definition of Sovereignty. *Millennium: Journal of International Studies*, 44(3), 305–320.

Latour, B. (2018a). *Down to Earth – Politics in the New Climatic Regime*. Cambridge: Polity Press.

Latour, B. (2018b). *A Clear Inversion of the End Times Schema*. From: www.bruno-latour.fr/sites/default/files/159-TIME%2bRAUCH-GB.pdf, viewed 30 June 2020.

Latour, B. (2019). *Issues with Engendering*. From: www.bruno-latour.fr/sites/default/files/167-TROUBLES-Engendering-GB.pdf, viewed 30 June 2020.

Latour, B. (2020a). *Seven Objections against Landing on Earth*. From www.bruno-latour.fr/sites/default/files/168-INTRO-CATALOG-semi-final-pdf.pdf, viewed 30 June 2020.

Latour, B (2020b). *Are You Ready to Extract Yourself from the Economy?* From: www.bruno-latour.fr/sites/default/files/downloads/P-205-ECONOMISATION-AOC-GB_1.pdf, viewed 30 June 2020.

Latour, B. and Chakrabarty, D. (2020). Conflicts of Planetary Proportions – a Conversation. From: www.bruno-latour.fr/sites/default/files/170-PLANETARY-PROPORTIONS.pdf, viewed 30 June 2020.

Levin, K., Cashore, B., Bernstein, S. and Auld, G. (2012). Overcoming the Tragedy of Super Wicked Problems: Constraining Our Future Selves to Ameliorate Global Climate Change. *Policy Sciences*, 45, 123–152.

Malm, A. (2020). *The Progress of This Storm. Nature and Society in a Warming World*. London: Verso.

Mann, M.E. (2020a). *Twitter Post,* 17 January. From: https://twitter.com/michaelemann.

Mann, M.E. (2020b). *Twitter Post,* 27 June. From: https://twitter.com/michaelemann.

Mann, M.E. (2021). *The New Climate War. The Fight to Take Back Our Planet*. New York: Public Affairs.

Marx, K. (1887). *Capital a Critique of Political Economy Volume I Book One: The Process of Production of Capital*. Moscow: Progress Publishers.

McKibben, B. (2010). *Earth: Making a Life on a Tough New Planet*. New York: Time Books.

Moore, J.W. (2015). *Capitalism in the Web of Life. Ecology and the Accumulation of Capital*. London: Verso.

Morris, D.W. (2019). A Human Tragedy? The Pace of Negative Global Change Exceeds Human Progress. *The Anthropocene Review*, 6(1–2), 55–70.

O'Malley, B. (2019). Networks of 7,000 Universities Declare Climate Emergency. *University World News*, 10 July. From: www.universityworldnews.com/post.php?story=20190710141435609, viewed 30 June 2020.

Olsson, G. (2007). *Abysmal. A Critique of Cartographic Reason*. Chicago: The University of Chicago Press.

Rankin, J. (2019). 'Our House Is on Fire': EU Parliament Declares Climate Emergency. *The Guardian*. From: www.theguardian.com/world/2019/nov/28/eu-parliament-declares-climate-emergency, viewed 30 June 2020.

Ripple, W.J., Wolf, C. Newsome, T.M., Barnard, P. Moomaw, W.R. and 11,258 scientist signatories from 153 countries (2020). World Scientists' Warning of a Climate Emergency. *BioScience*, 70(1), 8–12.

Rockström, J., Steffen, W., Noone, K. et al. (2009). A Safe Operating Space for Humanity. *Nature*, 461, 472–475. DOI: 10.1038/461472a.

Salamon, M.K. (2020). *Facing the Climate Emergency: How to Transform Yourself with Climate Truth*. Gabriola Island: New Society Publishers.

Sánchez-Bayo, F. and Wyckhuys, K.A.G. (2019). Worldwide Decline of the Entomofauna: A Review of Its Drivers. *Biological Conservation*, 232, 9–27.

Sloterdijk, P. (2016). *Foams: Spheres III*. Los Angeles: Semiotext(e).

Steffen, W. et al (2018). Trajectories of the Earth System in the Anthropocene. *PNAS Proceedings of the National Academy of Sciences of the Unites States of America*, 115(33), 8252–8259.

Swyngedouw, E. (2019). The Anthropo(obs)cene. In: *Keywords in Radical Geography: Antipode at 50, First Edition* (edited by The Antipode Editorial Collective). Hoboken: Wiley Blackwell, pp.253–258.

Thunberg, G. (2019). *No One Is Too Small to Make a Difference.* London: Penguin Random House.

UN (2019). United in Science. From: https://public.wmo.int/en/resources/united_in_science, viewed 30 June 2020.

Wallace-Wells, D. (2020). Global Warming Is Melting Our Sense of Time. *Intelligencer (Life after Warming)*, June 27, from: https://nymag.com/intelligencer/2020/06/global-warming-is-melting-our-sense-of-time.html, viewed 30 June 2020.

Williams, E.C. (2011). *Combined and Uneven Apocalypse: Luciferian Marxism.* Hants: Zero Books.

Xu, C., Kohler, T.A., Lenton, T.M., Svenning, J.-C. and Scheffer, M.M. (2020). Future of the human climate niche. *PNAS Proceedings of the National Academy of Sciences of the Unites States of America*, https://doi.org/10.1073/pnas.1910114117.

Zalasiewicz, J, Waters, C.N., Williams, M. and Summerhayes, C.P. (2019). *The Anthropocene as a geological time unit.* Cambridge: Cambridge University Press.

2 Do not go gentle into that good night

Contested narratives and political subjectivities in the Anthropocene

Carlos Tornel and Aapo Lunden

The Anthropocene's uncommitted potential

In 1947, when Dylan Thomas wrote the poem 'Do not go gentle into that good night', the idea of progress was rampant. In the poem, there is an undoubted approximation to the essence of modernity: 'rage against the dying of the light'. That is, to live in the fullest sense is presented as a narrative for both the limitations and capabilities of the human condition. The poem embodied a new meaning in Christopher Nolan's 2014 film *Interstellar* when Michael Cain's character reads the poem before launching the last manned mission to space and with it, the *last hope for humanity*. This interpretation of the lines in the film brings the poem into a new light, reflecting the human condition as a universalized 'we', who are in the midst of experiencing the Apocalypse. Nolan's film downplays the catastrophic elements. It does not linger in what originated the problem. Instead, it chooses to focus on the technological feats of human hubris as the redeemer. The film's argument revolves around the phrase 'We [human beings] were not meant to die on Earth, *we were meant to leave it*' (Nolan, 2014, our emphasis).

The plot of Interstellar serves as an excellent backdrop to address the Anthropocene and the politics of the Apocalypse. On the one hand, the Anthropocene is a condition of modernization, produced by accelerating the 'overheating' of the planet, not only due to carbon-based economy but also due to overheating of cultural representations, mobility, identity and natural ecosystems, built on progression without an aim for a credible and effective regulatory vision (Hylland, 2016). Moreover, the Anthropocene can be described as the end of separationist natures, due to the overflow of ecological conditions to already once purified socio-cultures of modernity (Latour, 1993). The shock over claiming socially constructed nature is largely a passé (Swyngedouw, 2011), as it is increasingly hard to 'maintain a separationist [...] [belief] that nature is a completely separate domain from social life' (Descola, 2013: 31).

Apocalyptic narratives, on the other hand, have been mobilized as an existential threat, requiring urgent if not immediate policy interventions.

DOI: 10.4324/9781003128854-3

Global environmental problems such as climate change, and biodiversity loss are framed as in need of extraordinary measures that, if not taken promptly, may lead to an apocalyptic and catastrophic scenario. However, while these threats are regarded as very much real, the emerging paradox is that the Apocalypse – represented as an external and antagonistic 'Other' or enemy – does not lead to exceptional measures, but rather micro and routine practices of risk management. The relentless invocation of the Apocalypse has led toward a piecemeal development and policies often led by technocrats, market-based instruments and technological innovations addressing the 'condition' of the Anthropocene in a post-political context (Methmann and Rothe, 2012).

Making of the apocalypse

Mainstream Anthropocene narratives pinpoint a particular use of the Apocalypse: mobilized as a prophylactic against alternative social and political imaginaries, ontologies and epistemologies seeking to disrupt the hegemonic notion of Geopower and promethean anthropocentrism (Rose, 2013). The Apocalypse is dreaded and invoked in three ways. First, it deploys a series of strategies and dispositives to reproduce capitalist economic order and absolute instrumentalization of nature, while simultaneously advocating for a profound agency of techno-managerial ecomodernism to reconfigure and retrofit neoliberal structures and subjectivities. Second, apocalyptic narratives are deployed as the source of a single 'origins story' of the Anthropocene, narrating a future of an imminently catastrophic Apocalypse. This is despite the fact that for other cosmologies, cultures or 'worlds' the end has already happened or is slowly taking place (Yusoff, 2013; Danowski and Viveiros de Castro, 2016). Finally, future apocalyptic narratives are presented in a context of dystopia and catastrophe where solutions are temporally and spatially detached from the existing crisis by individualizing responsibility, and shifting guilt from global actors to a universalized population. Here, the looming Apocalypse is presented as a threat that can only be constrained through an interrupted and unquestioned deployment of Geopower. However, there seems to be two main limitations to these kinds of thoughts around the Apocalypse. First, such notions are presented in a limited sense, enabling a linear temporality to depict events only occurring in the future. Second, the Apocalypse refers to the 'world' as a single problematic grand narrative. In other words, when addressing the Apocalypse as a looming, universal possibility, one still has to ask: whose world exactly is coming to an end? Whose world has already ended? Whose world will prevail, and how? Or, how should we position ourselves when facing the potential of multiple endings?

This chapter argues that Apocalyptic politics foreclose and disavow a series of ontologies and epistemologies, in which all other forms of knowledge and imaginaries are rendered silent or eliminated, while the Anthropocene is presented as the ultimate 'challenge' for Geopower. This is a challenge that 'we' must face – much in line with the plot of Interstellar – in order for 'us' not to die on Earth but nor leave it, but to make it manageable (again). Our purpose is to highlight how apotheotic managerialism, conceptualized as Geopower, constitutes an inequality and abolition of political ecologies and imaginations through post-political processes manifested in universalistic notions of the Apocalypse, agency and change. We claim that these narratives ultimately fail to recognize the emerging *pluriverse* of alternatives or what we, drawing form Delanty and Mota (2017), refer to as the Cosmopolocene, as a series of governing practices delinked from the traditional beliefs of development and modernity (Mignolo, 2013).

This chapter is divided as follows: in the second section, we review some of the keyways to conceptualize the Anthropocene as a grand narrative. In the third section, we turn to narratives dealing with how the Apocalypse is mobilized as a prophylactic seeking to depoliticize the grand narrative of the Anthropocene and sustain a universalized narrative of the *Earth-in-Peril*. Finally, in the fourth section, we explore alternative possibilities of political agency and emancipation in the Anthropocene.

Naming the Anthropocene

In this section, we describe three different approaches from the geological and social sciences around the concept of the Anthropocene as a geological epoch and a geopolitical event. Finally, we discuss the possibilities of planetary agency after the interruption of Gaia, and introduce the concept of Geopower to capture political ideologies of selective planetary submission and the challenges to these narratives.

The end of separationist nature

One of the central aspects of modernity is the idea of a separation between nature and culture (Latour, 1993). Claimed as a new epoch, the Anthropocene effectively abolishes this modern tradition. The separation between geological (Earth's) and biological (human) history that has originated since the nineteenth century presents nature as something external to history. However, as historian Dipesh Chakrabarty (2009) has argued, the sole presence of climate change demands a rewriting of history and globalization where the human and the non-human are reintegrated into the telling of 'deep history', a history in which the natural (geological) and human histories can be retold as a unified narrative leading us to the present geologic epoch: the Anthropocene. By canceling the separation between nature and culture, the Anthropocene forecloses the possibility to maintain any conception of external nature. This opens up an inquiry into political ecology of hybrid agency of nature and human societies. Thus, the Anthropocene needs to be understood as two particular, but not separate, conditions: as a *geological epoch* and as a *geopolitical event*.

The Anthropocene as an epoch and an event

In the geological epoch narrative, 'we' (the human species) have arrived at a new geological time where planetary conditions erupted as a direct causation of human activity on the planet: rapid acceleration of greenhouse gas emissions along with an unprecedented upsurge in energy use, biodiversity loss, air, water, land and atmospheric pollution, the intensification of biogeochemical cycles and the human impact on landscapes. These changes in the Earth's biosphere will leave an imprint of the 'human' presence for ages to come (Steffen et al., 2015). Thus, the geological Anthropocene presents a solid argument showcasing human influence and how it is affecting the parameters of the Earth system, accompanied by the potential to generate instability and even catastrophic shifts to the planet.

This epochal stance points to an understanding of the history of humans as overachievers in the planetary context (Hylland, 2016: 17–18). The geological aspect of the Anthropocene presents its thesis as an inescapable moment in space/time in which we have, inadvertently, 'ended up'. However, the idea that we have suddenly made sense of our collective impact and become aware of this condition through a collective awakening on the consequences of human actions is a fable (Bonneuil and Fressoz, 2016: xviii). The notion of a universal human *species* that is just becoming aware of the consequences of our collective actions presents a depoliticized narrative of the long history that constitutes what it means to live in the Anthropocene.

For Bonneuil and Fressoz (2016), the Anthropocene is not a condition but a geopolitical event, where human choices are imbued with planetary significance. For them, 'The Anthropocene is political inasmuch as it requires arbitrating between various conflicting human forces on the planet, between the footprints of different human groups (classes, nations), between different technological and industrial options, or between different ways of life and consumption' (2017: 26). In other words, the Anthropocene is not merely a natural or inescapable condition of the Earth, but rather a cultural model to constitute a new object of knowledge and order of governance (Delanty and Mota, 2017). Simply framing the Anthropocene as an inescapable geological epoch or as a cultural condition, risks sustaining the idea of (Western) human superiority by radically depoliticizing, off-staging and foreclosing any other epistemological and ontological understandings of the past, present and future.

The Good and the Bad Anthropocene

The interpretation that the Anthropocene has *arrived* as an inescapable spatio-temporal condition has been associated with two main currents of thought. First, what some authors have labeled the '*Good Anthropocene*' thesis (Dalby, 2015). This thesis is based on the faith on the ability of humans to 'collectively assert control over socio-ecological systems to good effect'

(Dryzek and Pickering, 2019). Advocates of the Good Anthropocene – more commonly associated with the Breakthrough Institute's 'The Ecomodernist Manifesto' (Asafu-Adjaye, et al. 2015, see also Shellenberg, 2020) – trust in human ingenuity, built-on large-scale deployment of, for example, genetically modified food, nuclear power, geoengineering and high-density urbanization. The main argument behind the Good Anthropocene is not that the human hubris and impact have gone too far, but rather that they have not gone far enough (Tsing, 2015). In this vein, humanity as a whole, would need to double down on the project of green capitalism, 'decoupling' (e.g. Bastani, 2019) to enable uninterrupted growth, development, urbanization and progress through ambitious techno-engineering projects with high material investment and big science (Asafu-Adjaye et al. 2015).

As such, the Good Anthropocene has exploited the idea of the Anthropocene becoming an umbrella term, a 'charismatic mega-category' (Malhi, 2017: 2) for all manner of environmental concerns and doomsday scenarios. Indeed, the warnings of a sixth mass extinction event (Kolbert, 2014), paralleled with the exposition of acceleration of climate science (IPCC, 2018) and the transgression of planetary boundaries (Rockström, et al. 2009; Steffen et al. 2015), have been articulated around a shibboleth, an intellectual zeitgeist (Lorimer, 2017) putting catastrophism and apocalyptic imaginaries at the center. Here, the Anthropocenic crisis is elevated to a global humanitarian concern, through meticulously depoliticized imaginary: one that elevates the argument of a 'more benign retrofitted climate exhaust[ing] the horizon of our social and political aspirations and imaginations' (Swyngedouw, 2011: 264).

The '*Bad Anthropocene*' (Dalby, 2015) narrative is built around the concept of 'Planetary Boundaries' (Rockström et al, 2009) which diversely supports the idea of 'returning to a stable condition' and 'a safe operating space for humanity', or in other words, the idea of 'turning back the clock' to the selectively stable conditions of a 'Holocene 2.0'. These conditions, however, seem more and more elusive. Hence, the 'bad' Anthropocene is characterized by an inescapable condition requiring a thorough 'reconceptualization and imagination of the institutions, practices, social structures, worldview, principles and systems' (Dryzek and Pickering, 2019: 12).

In a similar vein, Christ (2016) refers to the Anthropocene as a discourse that actually excludes the possibility to challenge human rule. The gesture of naming an Epoch focusing on the 'Anthropos' exposes a weakness of this concept as it continues to place human beings at the center of the argument; it silences, obscures and nullifies the human and non-human others 'who do not (as has been repeatedly established in the Western canon) speak, possess meanings, experience perspectives, or have a vested interest in their own destinies' (Christ, 2016: 18). For Christ, the term 'Anthropocene' implies a poverty in our nomenclature, limiting us to look beyond the human and nature divide, and insists in placing humanity as an 'ingenious unruly species' embodying the problem and the solution.

Recompositing Gaia

The idea that agency is entirely social needs to be abandoned (Urry, 2011: 8), along with the idea that society exists outside the realm of nature (Delanty and Mota, 2017). This new material temporality pronounced by a deep history is marked by what Isabelle Stengers calls 'the intrusion of Gaia' into the sphere of the Anthropos (Danowski and Viveros de Castro, 2016). Following Bruno Latour's notion, Gaia can be understood as the 'new state of nature' (Latour, 2017a), one that forces us to reckon with the impossibility of a human techno-utopia or perfect freedom resulting from the Promethean mastery of the Earth (Danowski and Viveiros de Castro, 2016), and instead acts as a force or events that effectively interrupt the spatiotemporality of the Holocene, imposing on us the need to acknowledge the end of purely Human agency in the Anthropocene. The intrusion of Gaia turns concepts – like Sustainable Development (SD) – that seek to substitute nature for 'the environment' on their head by framing crises as a problem of risk management, exemplified by concepts like 'resilience', 'circularity', 'decoupling' and/or 'internalizing externalities', that ineffectively question the actual possibility of neoliberalized environments to actually achieve an equilibrium between the economy and the ecology. Instead, the presence of Gaia opens a possibility for the totality of Earth itself to become an object of experimentation through an all-encompassing subsumption of the state of nature in late capitalism, opening up a new form of governance that is concerned with both the management of life (the bios) and the Earth (the geos): Geopower (Luisetti, 2019: 347).

The 'intrusion' of Gaia (Stengers, 2015) not only marks the collapse of the modern idea of separation (between nature and culture) but problematizes narratives of (SD) and its reifications. Hence, the Anthropocene thesis goes beyond the SD paradigm which seeks to reconcile ecology with economics after decades of ignoring the debates of the limits of growth, instead seeking to institute a process of risk management and resilience, one in which the Earth/Nature is transformed into a repository of ecosystem services (Lohmann, 2019). Here, the environment understood as 'that which surrounds us, the place where humans went to extract resources, deposit waste, or even in certain places was to be left virgin' (Bonneuil and Fressoz, 2016: 20), becomes obsolete along with the siren calls of sustainability in which nature is presented as externalized and fortified. Hence, the intrusion of Gaia forces us to simultaneously reflect on the impossibility of maintaining the paradigm of SD and to question the way in which this interruption is to be acknowledged in the Anthropocene.

Geopower

Some scholars have focused on what Christophe Bonneuil and Jean-Baptiste Fressoz (2016) call 'the shock of the Anthropocene'. This approach argues that the geological character of the Anthropocene as an epoch (discourse) obscures and forecloses a series of narratives, cosmologies and imaginations that systematically present this epochal crisis as universally produced (Malm, 2017), proposing a single way forward to address the crisis: that of Geopower. Geopower can be understood as the extension of the knowledge-power and governance of life (the bios) as an object and a political project (biopower) (Foucault, 2007), to the geos (the Earth as a whole), where the Earth becomes 'an object of knowledge (geo-knowledge) and government (Geopower)' (Bonneuil and Fressoz, 2016: 88). Geopower advocates doubling down on human agency over the Earth's systems and a neoliberal continuation of the political, economic and social structures of reproduction, consumption and exploitation. In other words, to guarantee that 'the balance of power is preserved, and the dangers of circulation and environmental transformations are contained by geosecurity' (Luisetti, 2019: 351). Indeed, the Anthropocene has become a dispositif to justify and deploy more apotheotic Earth management techniques to 'sustain the unsustainable' (Blühdorn, 2007).

Geopower can be considered as a combination of technocratic environmental interventions, geo-knowledge of imperial ecologies, Earth System Sciences and geoengineering (Bonneuil and Fressoz, 2016: 87–90). Drawing on the notion of biopower (Foucault, 2007), Luisetti argues that Geopower can be understood not only as the control techniques of population, but as *extended management of both life and non-life*. Thus, Geopower 'shifts governmental techniques from populations and goods to species, energy flows and ecosystems, from political economy to imperial ecologies, from economics and biology to the Earth system' (Luisetti, 2019: 351). Through Geopower, the Earth becomes a neoliberal repository of 'ecosystem services' in which natures are structured around neoliberal environments enforcing mechanisms to manage and control the Earth. The Anthropocene is thus understood as a new era of total subsumption of the bios (life) and the geos (non-life) under late capitalism, relating to what Mark Fisher (2009) assertively called 'capitalist realism': a process by which capitalism became ideologically uncontested, creating a 'widespread sense that, not only is capitalism the only viable political and economic system, but also that it is now impossible even to imagine a coherent alternative to it' (Fisher, 2009: 2).

Geopower systematically employs apocalyptic predictions as a source of fear and political mobilization where civilization and life, as we know it, may come to an end, unless something is done to avert the crisis (Klein, 2014). This process, fueled by what E. Ann Kaplan (2015) calls 'climate pre-trauma', renders a crisis into a manageable problem and is deployed as a strategy to disavow any attempts to disrupt, perturb and/or reconfigure the existing political order (Swyngedouw, 2018). Namely, only the technological and market-based solutions, such as extensive deployment of nuclear energy or geoengineering to cut current carbon emissions, are understood as possible solutions to the crises.

Contaminated options

As the Anthropocene sets the stage for the deployment of Geopower through neoliberal representations of natures and proposals based on technological optimism and market-based solutions, an aesthetic consisting of the Earth-at-risk is produced and advocated as a political ontology of commodifiable fear (Luisetti, 2019). This process is essential to the Earth project of Geopower. However, as the interruption and presence of Gaia becomes more and more evident, the Anthropocene has left us with no other choice but to 'get acquainted with her' (Latour, 2017b: vii). This means that the project of Geopower cannot simply manage the crisis but has to learn how to 'deal with it', opening new possibilities of governance and knowledge. However, despite a series of emerging epistemologies and ontologies in the Anthropocene, Geopower reduces 'our' options into four particularly politically contaminated options: (a) a prioritized trust in market-based solutions, (b) a total reliability in technological utopias, (c) an individualization of environmental responsibility (e.g., 'footprints') and (d) a romantic turn to conscious consumerism 'as if the rift was made whole when a privileged few shop at the farmer's market' (Wark, 2016: xv).

These positions not only offer depoliticized narratives displacing the environment from the political, but in fact foreclose other alternatives and plural knowledges that are essential in the questioning of Geopower. These alternatives are presented as a form of what Lauren Berlant calls 'cruel optimism: the compromised conditions of possibility by which realization is discovered either to be impossible, sheer fantasy, or too possible, and toxic' (Berlant, 2011: 24s). Wark (2016) argues that cruel optimism has been crucial in depoliticizing other liberation struggles and given rise to a technocratic 'Carbon Liberation Front' (CLF). This is understood as the global redistribution of molecules of carbon, giving birth to a sort of scientists-cum-revolutionary subjects capable of transforming the Earth through a large-scale deployment of geoengineering technologies and doubling down on the structures of Green capitalism, subsuming the whole Earth into a process of management and control (Armiero and De Angelis, 2017).

To sum up, the Anthropocene has reinstated universal grand narratives by placing all humans face to face with a common problem.

> However, a unifying universalism bears its own caveats. Placing a universalized 'we' in the center of the Anthropocene thesis risks maintaining modernity's project of obscuring, foreclosing and eliminating subaltern ontologies and other forms of knowledge opposing the separation of nature and culture. It demands a radical transformation of social and ecological relations and recognizes other forms of agencies from the non-human.
>
> (Swyngedouw and Ernstson, 2019)

The following section argues that Geopower can be exemplified through the use of catastrophic and apocalyptic imaginaries, creating a socio-natural project seeking to manage the entirety of the Earth. The aesthetic and the spectacle of the Earth-in-Peril narratives that emerge from the doomsday scenarios brought forth by climate and environmental science, international negotiations and Hollywood disaster films are exploited by those advocating Geopower to normalize existing conditions of an actually existing and sustained unsustainability (Blühdorn, 2013), and depoliticizing the socio-ecological predicament by leaving it in the hands of the experts (Swyngedouw and Ernstson, 2019: 34). We argue that, through this aesthetic, and not necessarily through the deployment of actual social-natural interventions, Geopower is able to colonize the global imaginary, and legitimize a sense of dependency on further acceleration and techno-managerial utopias to maintain current industrial civilizations indefinitely and unchanged.

'Managing' the apocalypse

Narratives about the Apocalypse are not new. Neither is the idea that humans generate their own demise. Every major religion has a version of this story, and the ethnographic records document a variety of ways in which human cultures have imagined their end. Here, we are interested in the political aspects of the Apocalyptic imaginations, aesthetics and narratives of Geopower in current Western politics and their use for precluding spaces of dissensus and disagreement to exert control over the population and the Earth (the bios and the geos).

As described above, the narratives and aesthetics of Geopower around catastrophism have become a central aspect of managing and sustaining the project of a *total management* of Earth, while perpetuating universal myths around the Anthropocene as an equal condition of humanity as a whole. The Anthropocene and the subsequent crises are then seen as the 'collective awakening' of our *species*, to the consequences of 'our' actions and impacts on the planet. As Ellis argues:

> Perhaps the most popular interpretation of the Anthropocene [...] is a catastrophic, human induced shift in the Earth's functioning as a system. In this view, recognizing the Anthropocene is the same as acknowledging the serious global consequences of climate change, mass extinctions and other anthropogenic environmental changes [...] Yet the Anthropocene itself is a synthesis of existing evidence, and not a new source of evidence for these changes or their consequences.
>
> (2018: 130)

Following on this notion of catastrophism, we highlight the main characteristics of apocalyptic narratives proliferating in contemporary politics over the essence and risks of the Anthropocene. We divide these characteristics into four themes: (a) mobilized inflexibilities, (b) guiding narratives, (c) rendering governable and (d) apotheosis.

Mobilized inflexibilities

First, the Apocalypse is mobilized to legitimize and fortify Geopower. Prime examples, such as CO_2 commodification through pricing and market exchanges, rely on extensive deployment of managerial governance regimes around risk calculation and financial tradings (Swyngedouw, 2010: 220). Furthermore, the deployment of geoengineering projects associated with Promethean promises and 'cornucopian-technophile utopias' (Danowski and Viveiros de Castro, 2016) assures that the 'capitalo-parliamentary' order remains intact and unchallenged as the universal order (Swyngedouw and Ernstson, 2019). Similar to the plot of the movie *Interstellar*, solutions to the existing catastrophe are presented as an issue of 'figuring out the science'. The 'Eureka moment' in the film is enough to sustain the existing order and effectively exit the Earth, depoliticizing the highly uneven impacts associated with the environmental degradation brought by capitalism. Here, the Apocalypse becomes a way to manage, bureaucratize and sustain the promise of redemption, via deployment of a limited set of both human and more-than-human socio-ecological options. This includes mobilized inflexibilities such as 'smart', 'sustainable', 'resilient' and/or 'adaptive' modes (Swyngedouw, 2019) of nurturing an imaginary capable of assuming a global supervisory role in managing the world's local and global ecologies (Swyngedouw and Ernstson, 2019: 31).

Guiding narratives

Second, narratives of the Apocalypse present the public as a population without attributes, whereas scientists are presented as quasi-revolutionary subjects (Armeiro and De Angelis, 2017). A crucial aspect of narratives such as the 'Good Anthropocene' and Geopower is that they place transformative agency in the hands of a managerial elite: the scientists. Paul Crutzen's seminal article in which he originally advocates for the use of the term 'Anthropocene' points to this:

> To develop a world-wide accepted strategy leading to sustainability of ecosystems against human induced stress will be one of the great tasks of mankind, requiring intensive research efforts and wise application of the knowledge thus acquired... *An exciting, but also difficult and daunting task lies ahead of the global research and engineering community to guide mankind towards global, sustainable, environmental management.*
>
> (Crutzen, 2002: 23. Our emphasis)

Crutzen's argument echoes Brundtland's report 15 years before: 'the new reality', referring to Earth as a single entity, 'from which there is no escape, must be recognized – and managed' (WCED, 1987). This aspect, which is most essential to Geopower, is sustained by dividing humanity into two categories: first, into a 'silent majority' of the world population – which has become a geological agent without realizing it; and second, into a scientific community holding a monopoly position capable of guiding masses and prescribing what needs to be done (Bonneuil and Fressoz, 2016: 80). Here, the Apocalypse is announced as a means to depoliticize the narrative, in which 'we' – the uninformed, passive, population without attributes – are presented as a universal victim of the coming Apocalypse. Simultaneously, the origins of the crisis (i.e. capitalism) are readily acknowledged but quickly externalized to a fetishized entity, which becomes a stand-in for the totality of the problem (CO_2 for climate change, externalities for biodiversity), and thus the enemy that must be eliminated to restore balance and justice (Swyngedouw, 2018: 303). As Armeiro and De Angelis (2017) argue, 'The risk is to envision the Anthropocene as a space for villains and victims but not for revolutionaries' (2017: 347), or as a space of futile citizenship, distanced from the realms and capacities of individuals to perceive causes and impacts of capitalist modernity, 'leaving citizens nothing but gloomy asceticism…' (Beck, 2010: 263).

Rendering governable

Third, the Apocalypse enables an immunological-biopolitical fantasy, to sustain an illusion of security and protection against the coming end of times. Swyngedouw and Ernstson (2018) define the Anthropocene discourse as an 'immunological prophylactic against the threat of an irredeemable external and vengeful nature' (2019: 35). They use the notion of biopolitics and immunitary politics (Esposito, 2008, 2011) to present a framework in which the governance apparatus (i.e., the set of practices, rules, institutions and techno-managerial proceedings) work to create an imaginary sense of protection and sequestration. This process takes control of the risks, dangers and fragilities of individuals to make them live in a huxleyan, peaceful manner, while obscuring any form of dissensus (Swyngedouw and Ernstson, 2019: 35–37).

The fantasy of absolute immunization (against the external other – i.e., CO_2) enables an imaginary sense of protection and sequestration, one in which the fears of collapse are displaced into the terrain of the governable and the space of adaptation, mostly through market-based mechanisms and techno-managerial forms of Geopower. Despite the shock that emerges by experiencing the presence of Gaia, the demands for urgency of action (i.e. nature-in-risk) are quickly internalized and rendered governable to political projects against the 'Other', which in the process turns the immunized society into passive spectators while depoliticizing any form of disagreement with the established order (Swyngedouw and Ernstson, 2019).

Thus, the apocalyptic catastrophe is rendered governable through the spectacular signing of international environmental agreements and further commodification of nature (Lohmann, 2019) where that which had been externalized (Gaia) is repurposed to sustain the socio-political and environmental composition of late capitalism. These choreographed spectacles of 'common agency' act as a form of stress relieving therapy where the spectacle (i.e. the signing of the Paris Agreement) is used to replace the global environmental crisis to overcome the fears over a self-made Apocalypse and restore trust in high-level decision-making. This process, what Swyngedouw and Ernstson (2019) call the Anthropo-Obscene, 'offers the promise of a radical reinterpretation of the apocalyptic endgame, without change; it promises the crafting of a fully manageable, manicured and inhabitable Earth' (Swyngedouw and Ernstson, 2019: 39).

Apotheosis

Fourth, the Apocalypse is mobilized as a way of presenting capitalism, and all its crisis tendencies (Fraser, 2014) as an inescapable natural condition generated with the advent of the Anthropocene. Here, the Apocalypse is mobilized as a catalytic event, a phantasmagorical imaginary threatening capitalist modernity, only to be solved by more capitalism. The argument here is brought back to the accelerationist processes and solutions promoted by apotheotic ecomoderns, fully automated luxury communists (Bastani, 2019) and the 'cornucopian technophiles', arguing that 'the world that has already started must *finish ending*' (Danowski and Viveiros de Castro, 2014: 52, our emphasis).

Thus, the Apocalypse is mobilized as a process in which the crisis tendencies and boundary struggles (the spheres of social reproduction, nature and politics), constituting the institutionalized social order of capitalism (Fraser and Jaeggi, 2018), will eventually self-reform. Here, the techno-utopian dreams are capable of overcoming the limitations and contradictions that until now have been fundamental for the actual production of value and the cycles of accumulation of capital (Moore, 2015). Interstellar, and the dreams of space travel embodied in private companies like Space-X present a version of this narrative, one in which the expansion of capitalism is infinite and cannot be limited by the planetary boundaries. This is the ultimate and final desire of Geopower: to exert total control over the geo, a process that is no longer limited by labor power or any interruption and is capable of expanding and terraforming beyond Earth (Fisher, 2009: 18).

Narratives of a forceful overcoming

Mark Fisher's analysis of Alfonso Cuaron's *The Children of Men* provides a good metaphor to sum up our argument and question the dominant role of the apocalyptic aesthetics, and the expected spectacle of Apocalypse in late capitalist societies:

> The catastrophe in Children of Men is neither waiting down the road, nor has it already happened. Rather, it is being lived through. There

> is no punctual moment of disaster; the world doesn't end with a bang, (rather) it winks out, unravels, gradually falls apart.
>
> (2009: 2)

Thus, the Apocalypse is not a singularity or an event that can be averted, but it is rather more in tune with what Rob Nixon calls 'slow violence', a process where violence 'occurs gradually and out of sight, a violence of delayed destruction that is dispersed across time and space, an attritional violence that is typically not viewed as violence at all' (Nixon, 2011: 2).

Danowski and Viveiros de Castro (2016) argue for a multiple understanding of 'worlds at ends', and question the overarching and universalizing narratives of one world and one end that is implicit, not only in the Anthropocene thesis but also in the Apocalypse that comes with it. For them:

> [The] variety of ends and of worlds [...] all seem to express the same fundamental historical condition: it has been disclosed to us that things are changing fast and not for the good of human life 'as we know it' [...] and more crucially, we have no idea what to do about it. The Anthropocene is the Apocalypse, in both the etymological and eschatological sense.
>
> (2017: 22)

The World, perhaps, has not ended, but we are experiencing multiple *ends* through *the progress of this storm* (Malm, 2017) and hence our politics must recognize the multiple endings and conditions of precarity and uncertainty emerging from the ruins of neoliberal capitalism (Tsing, 2015). In the following section, we argue that politics in the Anthropocene must be an emancipatory project, one that questions the dominant nature of Geopower and opens spaces for the emergence of other epistemologies and ontologies ostracized in the current narratives of the Apocalypse.

New political subjectivities

Bonneuil and Fressoz (2016) set out to present different historical threads that seek to rethink the visions of the world and how to inhabit it. By revisiting different histories of the Anthropocene and drawing distance from the dominant 'geocratic' account of the Anthropocene, they (re)conceptualize the Anthropocene as (among others) an *Agnotocene* (a history of ignoring planetary boundaries), as a *Capitalocene* (an attempt at a joint history of capitalism and the Anthropocene) or, indeed, as a *Polemocene*: a history of resistance, contestation and survival through the *long durée* of capitalism, colonialism and patriarchy by looking at the unequal relations of power and knowledges (Bonneuil and Fressoz, 2016).

The Polemocene exposes the existence of, 'since the eighteen century,[1] an environmentalism of the poor fighting for social justice and environmental decency' (Ibid.: 251), where struggles against environmental pillage, the enclosure of the commons and the exploitation of nature through a series of successive small coups, imposed situations and normalized exceptions has defined our epoch. We draw on these struggles and their ontological and epistemological frameworks to identify political subjectivities allowing us to navigate the increasingly diminishing world of the Anthropocene and the post-political project of Geopower, anchored on the construction of a looming Apocalypse.

Politics with Gaia

The Polemocene exposes how a notion of a collective agent misplaces the responsibility and simultaneously produces a mystification of the role of geologization of the social (Delanty and Mota, 2017). For Delanty and Mota, the Anthropocene has become a vehicle, or an arena in: 'which the human world is re-imagined culturally and politically in terms of its relations with the Earth' (2017: 34). The reconceptualization of spatio-temporal relations requiring us to 'think with Gaia' – given the impossibility to ignore or subvert her – has substantial implications in how 'we' govern the Anthropocene as a singular cultural model and moral space (Hylland, 2016), and needs to recognize the multiplicity of ecologies of knowledges (and knowledges of ecology) (de Sousa Santos, 2014: 42) through what they call a 'Cosmopolocene': a process that recognizes the alternatives and the plurality of ways of governing and inhabiting the Anthropocene (Delanty and Mota, 2017: 34).

The quest of identifying a political subject or subjectivity in the Anthropocene has also given rise to a new set of ontological paradigms questioning the agency of a supposedly universal 'Anthropos', the proposed genesis and adequate naming of the Anthropocene (see, for example, Malm and Hornborg, 2014; Haraway, 2016; Moore, 2015; Tsing, 2015; Lorimer, 2017; Malm, 2017). Advocates have focused their interests in producing a 'myriad of temporalities and spatialities and a myriad of intra-active entities-in-assemblages- including the more-than-human, other-than-human, inhuman, and human-as-humus' (Haraway, 2016: 160).

Planetary precarity and struggles of the multitude

Anna Tsing (2015) argues that the notion of the Anthropocene diverts attention from the patchy landscapes, multiple temporalities, and shifting assemblages of humans and non-humans: the very stuff of collaborative survival (2015: 20). Indeed, by presenting the solutions toward a geoengineered future, the ways in which multiple forms of cooperation (both between humans and non-humans) are entangled around the ruined landscapes of capitalism and those that emerge from their interactions (Tsing, et al., 2017) are shunned by the dominant discourses of technological development and market effectiveness. Tsing calls for a recognition of precarity as the predominant condition of our times, one in which the modern ideas of progress and the linearity of time cannot (continue to) be presented as the only solutions to the crisis. The ecomodern policies and practices that carried us into the Anthropocene foster radical precarity and uncertainty, and only by learning to live-with(-in) these conditions can a political strategy emerge.

For Armiero and De Angelis (2017), the Anthropos in the Anthropocene is a misplaced subject. Drawing on the logic of the Capitalocene, they expose how the notion of a collective 'we' must be questioned and understood, not as a universalized humanity but as a multitude of subjects 'whose practices are outside of the value practices of capital, often in the shape of commons systems' (Armeiro and De Angelis, 2017: 348). Drawing on Naomi Klein's (2014) concept of Blockadia and on the empirical evidence presented by the Global Environmental Justice Atlas (EJOLT), for them, the Anthropos is constituted by the struggles of *the multitude*, where *the commoners* are understood as the 'social subjects who collectively control, direct and engage in the reproduction of the commons and for which the relation to capital may often be necessary but does not exhaust their social being and activity' (Armeiro and De Angelis, 2017: 349).

Therefore, this brings the political to a crossroads: it forces us to recognize the excluded pluralities of knowledge that have been foreclosed by the totalizing narrative of the Anthropocene, and the ways in which struggles over the *commons* and the *multitude* can be politically articulated to challenge the narratives of Geopower and the Anthropocene. However, in doing so, it risks presenting an uncritical critique where we are 'all' victims of modernity and the Anthropocene, where the task would be to simply learn how to live in the ruined landscapes that come after capitalism (Chandler and Reid, 2019). The task then is to identify the multiple subjectivities that emerge from the cooperation and interactions between the human and the non-human, the pluriversal understanding of relational ontologies (Escobar, 2016) that can critically expose the depoliticizing narratives of the Anthropocene and the project of Geopower.

Searching for pluriversal politics in the Anthropocene: Epistemologies of the South

The problematization of a common Anthropocene opens a discussion over 'indigenous ontologies'. These ontologies are similar to what Latour called *the Terrans*: those others (non-moderns) that have continuously rejected the colonial and modernizing projects of development in the *long durée* of capitalism (Danowski and Viveiros de Castro, 2016). The emergence of different narratives questioning the hegemony of concepts like growth in the Global North (Kallis et al., 2018) and development in the Global South (Sachs, 2010) have produced an epistemic-political field toward a pluriverse of alternatives (Demaria and Kothari, 2017), one that can be understood as a rainbow of cosmologies, knowledges and vital worlds (Escobar, 2018).

Engaging with indigenous ontologies has led some authors to argue that humanity must abandon modernity and instead shift to a project of 'becoming indigenous' (Latour, 2018). The project of becoming indigenous is a way in which *the moderns* can learn from those that have learned to live in an impoverished world, who either resist or, after undergoing modernization, decide to 'go back' (Danowski and Viveiros de Castro, 2016). Becoming indigenous is then a project of 'governing the imaginary of what it means to be an agent or actor in a world, which is held to demand new forms of adaptivity, resourcefulness and resilience' (Chandler and Reid, 2019: 2). While the main argument here is that there is a need to move away from the 'modernizing' and 'developmental' narratives that continue to animate the 'Good Anthropocene', the critique to such an endeavor is that it risks an understanding of the Anthropocene not as a world of loom and gloom and extinction, but rather as 'an invitation to speculatively explore other ways of being and knowing' (Ibid, 2019: 13). This is problematic because it risks ignoring the contemporary realities of indigenous peoples and erasing the histories of difference.

For Bonaventura de Sousa Santos, the question is not of actually becoming indigenous (understood as an identitarian essentialism), but of recognizing a plural way of thinking and knowing to diversify narratives and repositories of concrete utopias of other possible worlds (2018). He utilizes the framework of Epistemologies of the South (EoS) to represent 'the production and validation of knowledge anchored in the experiences of resistance of all those social groups that have systematically suffered injustice, oppression, and destruction caused by capitalism, colonialism, and patriarchy' (2018: 1). This framework is then understood as a way to allow the oppressed social groups to represent the world as their own and in their own terms. For Arturo Escobar, unearthing such epistemologies itself takes an ontological turn: 'a process of multiple knowledges, or epistemes, refer to multiple worlds, or ontologies'. Here, the focus on the territorial struggles of peasants and indigenous communities serves as heuristics providing the essential elements for thinking about the profound cultural and ecological transitions needed to face the interrelated crises of climate, food, energy, poverty and meaning; a process that is itself attuned to the needs of the Earth (2016: 21).

Emerging transition discourses

Escobar (2015, 2016) argues that the framework of EoS offers a path toward interactions of multiple sets of ontological possibilities. These look beyond a particular construction of a 'One-World World' (OWW) – that is, a reality made up of a single 'World'– where the current existential crisis (the Anthropocene) points toward a need for a transition away from the reductionist colonial/modern construction of this particular vision of the World. Escobar (2015) refers to the interactions between the emerging transition discourses

(TDs) in different geographies as the space for a transition away from the OWW perspective. Alternative political movements, such as degrowth in Europe and Buen Vivir in Latin America, have started to question the supremacy of growth and argue for the construction of convivial societies, through a radical reconceptualization of worlds beyond capitalism, development and growth (Brand, 2020). These TDs,

> posit a profound cultural, economic, and political transformation of dominant institutions and practices. By making visible the damaging effects of dominant models of social life [...] they direct our attention to the need to transform culture and economy, oftentimes in tandem with these communities where the regimes of the individual, separation and the market have not yet taken a complete hold on socio-natural life.
>
> (Escobar, 2015: 454)

A relevant aspect of these TDs is that they do not imply a regression. In other words, degrowth and Buen Vivir are not advocating for returning to a pre-colonial condition or to a 'state of nature', but instead seek to question the notion of linear singular progress and history, the obsession with economic growth, consumerism and the commodification of nature, displacing humans as the only endowed subject of political representation and the colonial/Cartesian separation of nature from culture (Chuji, Rengfo and Gudynas, 2019: 112). Bridging a dialogue between concepts like degrowth and Buen Vivir, as well as other transition discourses emerging from the Epistemologies of the South, offers a way to radically question the mobilization of the apocalyptic and depolitical narratives of Geopower. Consequently, this points toward a possible way forward of actually constituting a Cosmopolocene, that is, a radical unraveling of the capitalist, colonialist and patriarchal notions constituting a one-way notion of history, development, progress and growth, currently embodied by the Anthropocene monocultural project (Escobar, 2015). Hence, the main challenge that remains is how to mobilize a global solidarity movement that pushes for an epistemic decolonization that allows us to 'degrow into a pluriverse' (Escobar, 2015: 458).

Silvia Rivera Cusicanqui's (2020) notion of *the Ch'ixi* can be a good way to understand how these many worlds can cooperate and coexist instead of being reduced to a universal framing of the World (OWW), its end (The looming Apocalypse) and its way forward (Geopower). The Ch'ixi reflects the Aymara idea of *something that is and is not at the same time*. It is the logic of the included third. The Ch'ixi expresses parallel coexistence of multiple cultural differences that do not extinguish themselves but instead antagonize and complement each other (Rivera Cusicanqui, 2020: 66). Equivalent to a 'motley' society, the Ch'ixi recognizes the bricolage of identities, knowledges and practices and hence proposes a construction based on differences that complement and antagonize each other. In doing so, it radically opposes a notion of becoming indigenous, arguing that the resulting hybrid subjects could only lead to condescending forms of tolerance, superficial multiculturalism and epistemological and ontological homogeneity.

Proposals like the Ch'ixi modernity have begun to envision a third space where, instead of 'becoming indigenous', the recognition of these motley fragments of the subaltern and laboring classes can emerge as a space for active dialogue between the politically active and their multiple ontological understandings of the world (Gago, 2020). These processes are characterized by confronting the colonization of the imaginary and the construction of knowledge enabling a more dynamic, intricate and often antagonistic forms of politics that 'allows the domains of the local and the planetary to become interwoven but without ever becoming hybridized or fused' (Arboleda, 2020: 28).

Re-learning to die and live in the Anthropocene

Hence, a political subjectivity in the Anthropocene requires an open debate about the universalization of a collective 'we', a process that questions the role of scientists and experts as orchestrators of 'our' politics and the constitution of the political passivity that is deployed through Geopower. The emerging pluriverse of alternatives recaptures what has been rendered obscure by capitalist, colonial and heteropatriarchal forms of governance. The trouble that erupts with the intrusion of Gaia opens new discussions about the way we should be able to navigate and live in the Anthropocene (Scranton, 2015). As such, Gaia is simultaneously exigent (it cannot be ignored) but it cannot be managed or mastered either (Stengers, 2018: 138). Rather, it requires that we think with and learn to live with Gaia, or succumb to the geo-engineered future and the hyper-modernist project of Geopower. Perhaps Bruno Latour summarizes it best when he argues that in fact there are only two possible alternatives: to modernize or to ecologize (Latour, 2018: 46).

The ideas of a linear progress and time that are implicit in the idea of the Anthropocene and the project of Geopower's relentless pursuit of economic growth and efficiency have become almost ubiquitous in the contemporary world, inasmuch as they have become an end in themselves (Malghan, 2019). On the one hand, the proponents of a 'Good' Anthropocene backing up accelerationist, Promethean and cornucopian futures construct their utopias by mobilizing depoliticizing narratives linked to an Apocalyptic vision of an end that is forever postponed as a strategy for geo-security. While on the other, the proponent of relational ontologies (like becoming indigenous) risks producing an uncritical critique of the Anthropocene and the apocalyptic narratives of the ends of the worlds, which present the problem as a question of coping with the Anthropocene. Thus, questioning the hegemonic project of Geopower must be based in recognizing the multiple ecologies of knowledge (and the knowledge of ecologies) (De Sousa Santos, 2014) as well as the ways of learning and inhabiting in this pluriverse of alternatives (Paulson, 2018).

As we remarked before, our politics must not be based on the name that we choose for our present condition, or on passively waiting for the Apocalypse that is infinitely postponed, to sustain a deeply depoliticized project of sustainability and its intention to expand to a planetary (and perhaps interplanetary) form of Geopower (Swyngedouw, 2018; Luisetti, 2019). On the contrary, the political must be based on what Donna Haraway (2016) refers to as: 'staying with the trouble', a process that consists of recognizing multiple worlds and multiple endings, but at the same time challenges us to find new forms of cooperation that allows to inhabit the ruined landscapes of capitalism. However, doing so must also mean radically denouncing the spectacle and depoliticized notions of the good, geoengineered Anthropocene and interrupt the project of Geopower. Shifting attention to the heterogeneity and difference of the political demands that emerge from the grounded normativity of socio-ecological movements (Coulthard and Simpson, 2016) is a way of staying with the trouble and opening up spaces for (re) politicizing the Anthropocene.

Conclusion: inequalities of imaginations

This chapter has drawn from key conversations focusing on the depoliticizing aspects of the mobilization of apocalyptic narratives in the Anthropocene. By reviewing different explanatory narratives, we point toward planetary discourses employing foreclosing strategies that, in our opinion, are unfruitful due to their dependency on visions of techno-managerialism. The main task of this chapter is to understand how the narratives of the Anthropocene, both from the geological and social sciences, have been associated with particular Apocalyptic narratives and discourses, and mobilized through a series of dispositifs and technologies enabling concepts like sustainability, growth and development to colonize the imaginaries of the future, and hence eliminate disagreement against this universalizing project identified as Geopower.

Taking the Anthropocene as a statement of fact risks obscuring the long history of violence, accumulation and dispossession, created through environmental pillage that has constituted the project of colonization and modernity (Ranganathan and Doshi, 2019). At the same time, by arguing for a 'good' Anthropocene, we risk institutionalizing a depoliticizing narrative that ultimately places the role of scientists and experts-cum-revolutionaries that decides our politics, creating citizens as spectators that subscribe to the immuno-biopolitical fantasy of securitization and sequestration away from the threat of 'the Other', in this case, the apocalyptic revenge of nature as Gaia (Swyngedouw and Ernstson, 2019).

Maria Kaika (2019) argues that the task of academics and intellectuals is to take on the challenge of trying to win the war of the imaginary – to link concepts to contemporary demands and struggles over socio-environmental change. This entails a step away from separationist teleologies. A global singularity is again on the agenda. But this is not due to the geologist framing of environmental change into manageable units to reduce ontological security (Giddens, 1991) of the affluent and adaptative, but due to political ecologization of the human existence, loss of flexibility due to material and political practices that allowed sumptuous and costly separation in the first hand. Yet, there is still a latent risk that the Anthropocene and apocalyptic framings become 'capitalism's latest apology: 'as a new master narrative, and as politically vacuous and complacent as, say, sustainability' (Kaika, 2019).

The task ahead then is to identify inequalities of imaginations and the ways in which dreams of utopias and apocalyptic visions lead to a radical foreclosure of alternatives in the Anthropocene. As Armeiro (2018: 198) argues, the Anthropocene does not deal with difference and is not interested in the nuances of the social construction of science nor situatedness. The immuno-biopolitical visions that emerge from these fantasies of securitization against a fetishized other (i.e. CO_2, Nature or population growth) can be sustained through the use of Apocalyptic imaginaries of the future (Swyngedouw and Ernstson, 2019). However, with the emergence of Geo-power, the control of life and non-life into a global hegemonic governance project produces what could be referred to as an *Immuno-Geopolitical framework* where life can be securitized, not only through human ingenuity and Promethean promises but through geological and non-living features of the *Whole Earth* as well.

In sum, our task is to search for the political subject in the absence of a universalized 'we', 'our' or 'us' based on a pluriversality or a Cosmopolocene, emerging from the recognition and demand for incorporation of different epistemologies and ontologies, to recognize existing inequalities and multiple political ecologies. Creating radical egalitarian spaces includes the capability of imagining, not the end of the world, but the people who will be able to inhabit it, to stay with the troubled future that is left to them. This process, as Bonaventura de Santos Sousa's (2018) would argue, is not marked by lack of alternatives but by a lack of alternative thinking of alternatives.

Acknowledgements

Our thanks to Arun Rao for his revision and commentary of this chapter. We also thank the editors of this volume and fellow authors for their comments and suggestions in developing the arguments presented here.

Note

1 We argue for a much longer historical understanding of the Capitalocene, or the Eurocene placing its origin in the sixteenth century with the colonization of the Americas and the input of cheap nature into the capitalist socio-economic system. See Moore (2015); Grove, (2019).

References

Arboleda, M. (2020) *Planetary Mine: Territories of Extraction under Late Capitalism.* London and New York: Verso.

Armeiro, M. (2018) Of ghosts, waste, and the Anthropocene. In: H. Ernstson and E. Swyngedouw. Eds., *Urban Political Ecologies in the Anthropocene. Interruptions and Possibilities.* London and New York: Routledge, 184–202.

Armerio, M. and De Angelis, M. (2017) Anthropocene: Victims, narrators, and revolutionaries. *The South Atlantic Quarterly* 116(2): 345–362.

Asafu-Adjaye, J., Christopher, F., Rachel, P., Linus, B., David, K., Jayashree, R., Martin, L., Stewart, B., Sagoff, M., Brook, B., Lynas, M., Shellenberger, M., Defries, R., Nordhaus, T., Stone, R., Ellis, E., Pielke, R. and Teague, P. (2015) *An Ecomodernist Manifesto.* Available at: www.ecomodernism.org. [Accessed on 08 August, 2020].

Bastani, A. (2019) *Fully Automated Luxury Communism.* London and New York: Verso Books.

Beck, U. (2010) Climate for change, or how to create a green modernity*? Theory, Culture & Society* 27(2–3): 254–266.

Berlant, L. (2011) *Cruel Optimism.* Durham and London: Duke University Press.

Blühdorn, I. (2007) Sustaining the unsustainable: Symbolic politics and the politics of simulation. *Environmental Politics* 16(2): 251–275.

Blühdorn, I. (2013) The governance of unsustainability: Ecology and democracy after the post-democratic turn. *Environmental Politics* 22(1): 16–36.

Bonneuil, C. and Fressoz, J. (2016) *The Shock of the Anthropocene: The Earth, History and Us.* London: Verso.

Brand, U. (2020) Post-extractivism: Against the exploitation of natural resources. In: C. Burkhart, M. Schmelzer and N. Treu. Eds., *Degrowth in Movements. Exploring Pathways for Transformation.* Winchester and Washington: Zero Books, 244–257.

Chandler, D. and Reid, J. (2019) *Becoming Indigenous: Governing Imaginaries in the Anthropocene.* London: Rowman & Littlefield, Ltd.

Chakrabarty, D. (2009) The climate of history: Four theses. *Critical Inquiry* 35(2): 197–222.

Christ, E. (2016) On the poverty of the nomenclature. In: J.W. Moore. Ed., *Anthropocene or Capitalocene. Nature, History and the Crisis of Capitalism.* Oakland: PMPress, 14–33.

Chuji, M., Rengfo, G. and Gudynas, E. (2019) Buen vivir. In: A. Kothari, A. Salleh, A. Escobar, F. de Maria, and A. Acosta. Eds., *Pluriverse: A Post Development Dictionary.* Delhi: Tulika Books, 111–113.

Coulthard, G. and Simpson, B. (2016) *Grounded Normativity/Place-Based Solidarity. American Quarterly* 68(2): 249–255.

Crutzen, P. (2002) Geology of mankind: The Anthropocene. *Nature* 415: 23.

Dalby, S. (2015) Framing the Anthropocene: The good, the bad and the ugly. *The Anthropocene Review* 3(1): 33–51.

Danowski, D. and Viveiros de Castro, E. (2016) *The Ends of the World.* Cambridge: Polity Press.

De Maria, F. and Kothari, A. (2017) The post-development dictionary agenda: Paths to the pluriverse. *Third World Quarterly* 38(12): 2588–2599.

De Sousa Santos, B. (2014) *Epistemologies of the South: Justice against Epistemicide.* Boulder: Paradigm Publishers.

De Sousa Santos, B. (2018) *The End of the Cognitive Empire. The Coming of Age of Epistemologies of the South*. Durham and London: Duke University Press.

Delanty, G. and Mota, A. (2017) Governing the Anthropocene: Agency, governance, knowledge. *European Journal of Social Theory* 20(1): 9–38.

Descola, P. (2013) *Beyond Nature and Culture*. Chicago: University of Chicago Press.

Dryzek, J.S. and Pickering, J. (2019) *The Politics of the Anthropocene*. Oxford: Oxford University Press.

Ellis, E. (2018) *Anthropocene. A Very Short Introduction*. Oxford: Oxford University Press.

Escobar, A. (2015) Degrowth, postdevelopment, and transitions: A preliminary conversation. *Sustainability Science* 10(3): 451–462.

Escobar, A. (2016) Thinking -feeling with the Earth: Territorial struggles and the ontological dimension of the epistemologies of the South. *Revista de Antropología Iberoamericana* 11(1): 11–32.

Escobar, A. (2018) *Designs for the Pluriverse. Radical Interdependence, Autonomy and the Making of Worlds*. Durham and London: Duke University Press.

Esposito, R. (2008) *Bios*. Minneapolis: University of Minnesota Press.

Esposito, R. (2011) *Immunitas*. Cambridge: Polity Press.

Fisher, M. (2009) *Capitalist Realism. Is There No Alternative?* Hampshire: Zero Books.

Foucault, M. (2007) *Security, Territory, Population: Lectures at the Collège de France, 1977–1978*. London: Palgrave Macmillan.

Fraser, N. (2014) Behind Marx's hidden abode: For an expanded conception of capitalism. *New Left Review* 8(6): 55–72.

Fraser, N. and Jaeggi, R. (2018) *Capitalism: A Conversation in Critical Theory*. Cambridge: Polity Press.

Gago, V. (2020) Introduction. The Silvia Rivera Cusicanqui principle: The rebellion of thought. In: Ch'ixinakax utxiwa. Ed., *On Practices and Discourses of Decolonization by S. Rivera Cusicanqui*. Cambridge: Polity Press, vii–xxxiii.

Giddens, A. (1991) *Modernity and Self-Identity: Self and Society in the Late Modern Age*. Stanford: Stanford University Press.

Grove, J.V. (2019) *Savage Ecology: War and Geopolitics at the End of the World*. Durham and London: Duke University Press.

Haraway, D. (2016) *Staying with the Trouble: Making Kin in the Chthulucene*. Durham: Duke University Press.

Hylland, E.T. (2016) *Overheating: An Anthropology of Accelerated Change*. London: Pluto Press.

Intergovernmental Panel on Climate Change (IPCC) (2018) *Special Report on Global Warming of 1.5 °C*. Available at: www.ipcc.ch/sr15/. [Accessed 08 August, 2020].

Kaika, M. (2019) Reclaiming a scholarship of presence: Building alternative socio-environmental imaginaries. In: H. Ernstson and E. Swyngedouw. Eds., *Urban Political Ecologies in the Anthropocene. Interruptions and Possibilities*. London and New York: Routledge, 239–252.

Kallis, G., Kostakis, V., Lange, S., Muraca, B., Paulson, S. and Schmelzer, M. (2018) Research on degrowth. *Annual Review of Environment and Resources* 43: 1–26.

Kaplan, E.A. (2015) *Climate Trauma: Foreseeing the Future in Dystopian Film and Fiction*. New Jersey: Rutgers University Press.

Klein, N. (2014) *This Changes Everything: Capitalism vs. the Climate*. New York: Simon & Schuster.

Kolbert, E. (2014) *The Sixth Extinction: An Unnatural History.* New York: Henry Holt and Company.

Latour, B. (1993) *We Have Never Been Modern.* Cambridge: Harvard University Press.

Latour, B. (2017a) *Facing Gaia. Eight Lectures on the New Climatic Regime.* Cambridge: Polity Press.

Latour, B. (2017b) Forward. In: D. Danowski and E. Viveiros de Castro. Eds., *The Ends of the World.* Cambridge: Polity Press, vii–ix.

Latour, B. (2018) *Down to Earth. Politics in the New Climatic Regime.* Cambridge: Polity Press.

Lohmann, L. (2019) Ecosystem services trading. In: A. Kothari, A. Salleh, A. Escobar, F. de Maria and A. Acosta. Eds., *Pluriverse: A Post Development Dictionary.* Delhi: Tulika Books, 47–50.

Lorimer, J. (2017) The Anthropo-scene: A guide for the perplexed. *Social Studies of Science* 47(1): 117–142.

Luisetti, F. (2019) Geopower: On the states of nature of late capitalism. *European Journal of Social Theory*, 22(3): 342–363.

Malghan, D. (2019) Efficiency. In: A. Kothari, A. Salleh, A. Escobar, F. de Maria and A. Acosta. Eds., *Pluriverse: A Post Development Dictionary.* Delhi: Tulika Books, 50–53.

Malhi, Y. (2017) The concept of the Anthropocene. *Annual Review of Environment and Resources*, 42: 77–104.

Malm, A. (2017) *The Progress of this Storm. Nature and Society in a Warming World.* New York and London: Verso.

Malm, A. and Hornborg, A. (2014) The geology of mankind? A critique of the Anthropocene narrative. *Anthropocene Review* 1(1): 62–69.

Methmann, C. and Rothe, D. (2012) Politics for the day after tomorrow: The logic of apocalypse in global climate politics. *Security Dialogue* 43(4): 323–344.

Mignolo, W. (2013) *On Pluriversality*, waltermignolo.com. Available at: http://waltermignolo.com/on-pluriversality/. [Accessed 08 August, 2020].

Moore, J.W. (2015) *Capitalism in the Web of Life. Ecology and the Accumulation of Capital.* London and New York: Verso Books.

Nixon, R. (2011) *Slow Violence and the Environmentalism of the Poor.* Cambridge and London: Harvard University Press.

Nolan, C. (Director) (2014) *Interstellar* [Motion Picture]. United States: Paramount Pictures and Warner Bros. Pictures.

Paulson, S. (2018) Pluriversal learning: Pathways toward a world of many worlds Nordia Geographical Publications. *NGP Yearbook 2018: Affirmative Political Ecology* 47(5): 85–109.

Ranganathan, M. and Doshi, S. (2019) Political ecologies of dispossession and anticorruption: A radical politics for the Anthropocene? In: H. Ernstson and E. Swyngedouw. Eds., *Urban Political Ecologies in the Anthropocene. Interruptions and Possibilities.* London and New York: Routledge, 91–110.

Rivera Cusicanqui, S. (2020) *Ch'ixinakax utxiwa. On Practices and Discourses of Decolonization.* Cambridge: Polity Press.

Rockström, J., Steffen, W., Noone, K., Persson, Å., Chapin, III, F.S., Lambin, E., Lenton, T.M., Scheffer, M., Folke, C., Schellnhuber, H., Nykvist, B., De Wit, C.A., Hughes, T., van der Leeuw, S., Rodhe, H., Sörlin, S., Snyder, P.K., Costanza, R., Svedin, U., Falkenmark, M., Karlberg, L., Corell, R.W., Fabry, V.J., Hansen, J., Walker, B., Liverman, D., Richardson, K., Crutzen, P. and Foley, J. (2009) Planetary boundaries: Exploring the safe operating space for humanity. *Ecology and Society* 14(2). www.ecologyandsociety.org/vol14/iss2/art32/

Rose, D.B. (2013) Anthropocene noir. *Arena Journal* 41/42: 206–219.
Sachs, W. (2010) Introduction. In: W. Sachs. Ed., *The Development Dictionary. A Guide to Knowledge as Power.* Second Edition. London and New York: Zed Books, xv–xx.
Scranton, R. (2015) *Learning to Die in the Anthropocene. Reflections on the End of Civilization.* San Francisco: City Lights Publishers.
Shellenberger, M. (2020). *Apocalypse Never: Why Environmental Alarmism Hurts Us All.* New York: Harper.
Steffen, W., Richardson, K., Rockström, J., Cornell, S.R., Fetzer, I., Bennett, E.M., Biggs, R., Carpenter, S.R., de Vries, W., de Wit, C.A., Folke, C., Gerten, D., Heinke, J., Mace, G.M., Persson, L.M., Ramanathan, V., Reyers, B. and Sörlin, S. (2015) Planetary boundaries: Guiding human development on a changing planet. *Science* 347(6223). Available at: http://science.sciencemag.org/content/347/6223/1259855. [Accessed 08 August, 2020].
Stengers, I. (2015) *In Catastrophic Times: Resisting the Coming Barbarism.* London: Open Humanities Press.
Stengers, I. (2018) *Another Science Is Possible. A Manifesto for Slow Science* Cambridge: Polity Press.
Swyngedouw, E. (2010) Apocalypse forever? *Theory, Culture & Society*, 27(2–3): 213–232.
Swyngedouw, E. (2011) Depoliticized environments: The end of nature, climate change, and the postpolitical condition. *Royal Institute of Philosophy Supplement* 69: 253–274.
Swyngedouw, E. (2013) Apocalypse now! Fear and doomsday pleasures. *Capitalism Nature Socialism* 24(1): 9–18.
Swyngedouw, E. (2018) Co2 as neoliberal fetish: The love of crisis and the depoliticized immuno-biopolitics of climate change governance. In: D. Cahill, M. Cooper and M. Konings. Eds., *The Sage Handbook of Neoliberalism.* London: Sage Publications, 295–307.
Swyngedouw, E. (2019) The Anthropo(Obs)cene. In: Antipode Editorial Collective. Ed., *Keywords in Radical Geography: Antipode at 50.* Hoboken: John Wiley & Sons, 253–258.
Swyngedouw, E. and Ernstson, H. (2019) O Tempora! O Mores! Interrupting the Anthropo-obScene. In: H. Ernstson and E. Swyngedouw. Eds., *Urban Political Ecologies in the Anthropocene: Interruptions and Possibilities.* London and New York: Routledge, 25–47.
Tsing, A. (2015) *The Mushroom at the End of the World: On the Possibility of Life in Capitalist Ruins.* Princeton: Princeton University Press.
Tsing, A.L., Bubandt, N., Gan, E., and Swanson, H.A. (2017) *Arts of Living on a Damaged Planet: Ghosts and Monsters of the Anthropocene.* Minnesota: University of Minnesota Press.
Urry, J. (2011) *Climate Change and Society.* Cambridge: Polity Press.
Wark, M. (2016) *Molecular Red: Theory for the Anthropocene.* London and New York: Verso Books.
World Commission on Environment and Development (WCED) (1987) *Our Common Future.* Oxford: Oxford University Press.
Yusoff, K. (2013) Geologic life: Prehistory, climate, futures in the Anthropocene. *Environment and Planning D: Society and Space* 31(5): 779–795.

3 The end of worlding

Indigenous cosmologies in the Anthropocene

Mariana Reyes-Carranza

> Another end of the world is possible.
>
> (message written on a wall in Santiago, Chile)

Unquestionably, the Anthropocene has become the most influential concept in environmental studies in recent decades (Moore, 2016, p. 2). The flow of colloquiums, publications, projects, performances, and installations about the Anthropocene evidences its success in the academic, artistic, and media worlds. However, its acceptance as a signifier of the environmental crisis has resulted in a proliferation of disagreements and a lack of consensus on the effectiveness of its use. For instance, in analysing the Anthropocene's apocalyptic rhetoric, scholars and commentators have denounced the term as a source of political paralysis (Cook and Balayannis, 2015; Haraway, 2016; Swyngedouw, 2013). Likewise, feminist, Indigenous, and subaltern voices have made urgent calls to decolonise or indigenise the Anthropocene, that is, to challenge its Eurocentric bias and to make space for the perspectives of colonised, racialised, and *othered* peoples (Davis and Todd, 2017; Morrison, 2015; Simpson, 2018; Todd, 2015; Whyte, 2017, 2018).

Against this backdrop, this chapter discusses Indigenous[1] cosmologies in a time of climate change and mounting ecological catastrophe. Special focus is placed on Amerindian stories, notably the mythological universe of the Yanomami people, an ethnic group living in the Amazon rainforest, on the border between Venezuela and Brazil. In particular, I offer an in-depth exploration of *The Falling Sky*, a book jointly written by shaman Davi Kopenawa and French anthropologist Bruce Albert. Described as an Indigenous world-ending story, *The Falling Sky* has triggered numerous theorisations that link it to the thesis of the Anthropocene (e.g. Arregui, 2019; Danowski and Viveiros de Castro, 2016; Szerszynski, 2017). Inspired by these valuable contributions, I present an analysis of the spatial dimensions of the Anthropocene, that is, the ways in which the effects of climate and environmental breakdown manifest themselves in specific, situated geographies. To that end, I engage with two geographical locations: the

DOI: 10.4324/9781003128854-4

Yanomami lands, where most of the narrated events in *The Falling Sky* take place, and the Museum of Tomorrow in Rio de Janeiro, on the Brazilian Atlantic coast. Whilst culturally and geographically distant, the museum and the Yanomami territory serve as points of departure for an exploration of the ways in which the "end of the world" is understood in distinct cultural and political contexts.

This chapter is divided into four sections. The first one introduces Yanomami mythological narratives of the end of the world, and their affinities with scientific discourses of climate change and ecological degradation. The second section focuses on the Museum of Tomorrow, a cultural project whose use of apocalyptic imagery attempts to raise awareness of the imperilled future of humanity and the planet. The third section interrogates the idea of the apocalypse as an embodied, situated, and contextualised experience of the climate and ecological crises. Lastly, the chapter concludes with suggestions on the role non-Indigenous scholars can take in uplifting Indigenous lived experiences and epistemes as a means of conjuring liveable futures beyond the destruction of the Anthropocene.

The Falling Sky

In 1975, French anthropologist Bruce Albert arrived in the remote lands of the Amazon rainforest. His ethnographical curiosity had led him to a territory inhabited by the Yanomami people, an Amerindian group living in the interfluvial lands between the Orinoco and Parima Rivers. Believed to have remained in isolation for millennia, the Yanomami had developed specific genetic and linguistic features, distinct from other Indigenous groups in the surrounding areas (Albert, 2010). Whilst they once occupied a vast territory, the arrival of settlers from the eighteenth century onwards caused mass displacement among the Yanomami people and the subsequent emptying of their original lands. Moreover, sporadic encounters between Yanomami and explorers, alongside misleading anthropological accounts, resulted in stereotyped and racist portrayals of the Yanomami people as a bloodthirsty tribe (e.g. Chagnon, 1988).

In the 1970s, a major project threatened to severely affect the Yanomami population. The Brazilian military government of the time ordered the construction of a highway connecting to the north of the country at the expense of the well-preserved Amazonian ecosystem. By the time Albert arrived in the area, construction of the road had not only destroyed some Indigenous villages and hunting areas, but led to many deaths caused by the introduction of diseases to which the Indigenous people had no immunity (Albert, 2010).

Following his arrival, the young anthropologist settled in a village in the proximity of the Catrimani River. There, Albert gained a basic understanding of the Yanomami language, as well as their social and ritual organisation. A few years later, Albert met Davi Kopenawa, a young shaman and interpreter for the Brazilian government. They became close friends and partners in an emergent political struggle for the defence of Indigenous rights, and the legal demarcation of Yanomami lands. Their involvement in such ethno-political mobilisation resulted in the co-authored book *The Falling Sky*, a text that affirms the main strands of Yanomami cosmological thought, following the shamanic practice of Davi Kopenawa.

Broadly speaking, *The Falling Sky* describes Kopenawa's shamanic knowledge, informed by the dreams and visions induced by the hallucinogenic plant known as *yãkoana*. The plant's use allows shamans to come into contact with the powerful *xapirí* (spirits beings), who are understood as mirror images of any entity of the world. As Kopenawa and Albert (2013, p. 490) explain,

> [a]ny existing being has an "image" from the original times, an image which shamans can 'call,' 'bring down,' and 'make dance' as an 'auxiliary spirit.' These primordial image-beings ('spirits') are described as miniscule humanoids wearing extremely bright, colorful feather ornaments and body paint.

When a shaman uses the *yãkoana* plant, he becomes a spirit himself, calling for other spirits to come down from the sky. This inter-world communication guides shamans through several endeavours, from healing people to "calling" the rain or halting bad weather. As such, shamanism informs a type of relational encounter between the different entities – both human and non-human - that populate the forest, turning shamans into mediators between beings (Viveiros de Castro, 2009).

Moreover, as its title suggests, *The Falling Sky* alludes to one of the Yanomami's most significant shamanic prophecies: the advent of a "gold" smoke, the death of all shamans, and the eventual falling of the firmament. According to this prediction, the forest ground contains a dangerous epidemic smoke called *xawara*. If released from the ground, the smoke will expand through the air and reach the "sky chest," sickening both the people's lungs and the sky, and eventually making it fall. Despite shamanic efforts to sustain the sky, the epidemic smoke accumulation will ultimately tear it down. In Kopenawa and Albert's (2013, p. xvii) words,

> [shamans] will no longer be able to hold back the evil beings who will turn the forest to chaos. We will die one after the other, the white people as well as us. All the shamans will finally perish. Then, if none of them survives to hold it up, the sky will fall.

According to Yanomami cosmology, the evil smoke comes out of the oil and metal deposits that lie under the forest. These substances are the remains of an earlier sky that fell apart in time immemorial. Hence, the minerals and other resources that comprise the Earth's crust were once part of a former sky. Moreover, such materials are meant to remain underground by order of the demiurge *Omama*:

> The words of *Omama*'s image teach us to beware of gold and the other minerals. They are unknown and dangerous evil things that only bring disease and death. When it is still in the form of a stone, gold is a living being. It only dies by melting in the fire, when its blood evaporates in the big pots of white people's factories. As it dies, it lets off the dangerous heat of its breath, which we call *oru a wakëxi*, the gold smoke. All the minerals do this when they are burned. This is why the fumes from metal, motor oil, tools, pots, and all the objects made by white people mix together and spread through the cities. These vapours are hot, thick, and yellowish like gas. They stick to hair and clothing. They get into eyes and invade chests. This poison fouls white people's bodies without their knowing. Then this evil smoke drifts far into the distance, and when it reaches the forest, it tears up our throat and devours our lungs. It burns us with its fever and makes us cough endlessly, then weakens our flesh before killing us.
>
> (Kopenawa and Albert, 2013, p. 288)

The story of the falling sky speaks of a past mythical event that simultaneously prefigures the destiny of our contemporary world. Just as the sky fell before destroying a previous form of humanity, it will fall again, this time due to the deadly emissions produced by white people's[2] practices of mineral extraction. Reading *The Falling Sky* against today's ecological catastrophe, Szerszynski (2017) argues that Yanomami's mythical interpretation of unearthed metals denounces the consumerist and energy-addicted practices of Western society. Hence, it could be said that Yanomami storytelling sends a cautionary message about the damaging effects of the extractive and fossil fuel industries. Put another way, Yanomami cosmology denounces human-induced pollution caused by the economic greed of the "people of merchandise," a people blindly obsessed with obtaining oil and metals for the manufacture of objects and machines. As Szerszynski (2017, p. 266) summarises, "[*The Falling Sky*] is a territorialisation of multiple forms of Anthropocene violence." In it, "the creation of desire for consumer goods, the mobilisation of long-buried geological resources and the disruption of the atmospheric commons are coded onto the back of the forest and the chest of the current sky" (Szerszynski, 2017, p. 266).

Yet another thought-provoking observation about Yanomami eschatology concerns its understanding of the "end of the world" as the ultimate obliteration of the forest. As shown in *The Falling Sky*, the space over which the sky threatens to fall is *urihi a* (the forest), a term Kopenawa uses to translate the Portuguese word *mundo* (world). Put simply, the forest is the world, and the world is the forest (Viveiros de Castro, 2019). Therefore, the end of the world is the end of the forest:

> The forest is alive. It can only die if white people persist in destroying it. If they succeed, the rivers will disappear underground, the soil will crumble, the trees will shrivel up, and the stones will crack in the heat. The dried-up earth will become empty and silent. The *xapiri* spirits who come down from the mountains [...] will escape far away. Their shaman fathers will no longer be able to call them and make them dance to protect us. They will be powerless to repel the epidemic fumes which devour us.
>
> (Kopenawa and Albert, 2013, p. xvii)

Drawing parallels between *The Falling Sky* and the idea of the Anthropocene, Brazilian anthropologist Eduardo Viveiros de Castro has discussed the possibilities of translating the ideas in the text into the language of Western philosophical discourse. As he argues (2019, p. 9),

> Amerindian ways of thought, independently of the quantity and quality of "objective truths" they give access to, allow at the very minimum a commerce with reality that is less arrogant, less imperial, less destructive than the one facilitated, not to say enforced, by our cosmological vulgate.

Following Viveiros de Castro's reading of *The Falling Sky* as a prophetic discourse on the Anthropocene, I contend that Yanomami storytelling, rather than depicting the end of the world, speaks of what I will term here the "end of worlding."

Instead of a single disastrous event, for Yanomami people, the end of the world is an ever-present danger. In their understanding, "multiple temporalities are generated within the extended body of the planet so that multiple world-beginnings and world-endings are internal to the process of planetary self-organisation" (Szerszynski, 2019, p. 204). In this line of thinking, Yanomami people know that for the sky not to fall, the world must be actively maintained through ritual practices and inter-species gifts. Thus, it becomes imperative to maintain the generative conditions of the world. This means not only preserving traditional knowledge and ritual practice, but maintaining the traditional lifeways that make such knowledge intelligible (Moeller, 2018 as cited in Szerszynski, 2019).

Furthermore, the Yanomami perception of the world differs broadly from Western-centric ontologies, which generally conceive of the world as a unity that will go on existing whatever humans do (Szerszynski, 2019). Indeed, Kopenawa claims that *urihi a* (the forest) is a whole – a non-fractionable entity (Kopenawa and Albert, 2013, p. 397). Nonetheless, this entity is simultaneously composed of a complex entanglement of beings: the human and non-human, spirits, and the land. If indeed a world is not only a planet but "a set of relations, obligations, and mutual dependencies" (Szerszynski, 2019, p. 206), the end of the world is the termination of this set of connections, that is, the end of world-making practices, or what I term here "the end of worlding."

By the "end of worlding," I refer to the challenges faced by Yanomami people – and other Amerindian groups – in sustaining their existence and preserving their ancestral practices and relations to their kin and companion species. Surrounded by the might of extractive and capitalist lifeways, Yanomami people physically and ideologically resist the shockwaves of a rapidly changing environment. Living on the frontline of environmental destruction, this people's most pressing concerns do not revolve around the Anthropocene's nomenclature and chronology, but around the vital need to resist growing environmental decline whilst maintaining their ancestral forms of existence. As Manuel Tironi (2019) notes, many contemporary societies are not politically concerned with discussing the scientific validity of the Anthropocene as a geological era. Instead, and in light of the specific vulnerabilities and possibilities of their places, these groups speculate about what kinds of ethics, practices, and pedagogies are needed to live through and cope with the destruction of the Anthropocene.

Following this, in the next section I consider how apocalyptic narratives of the Anthropocene, instead of inspiring reflection on the contemporary environmental crisis, enhance an end-of-time paralysis. Ironically, these doomsday scenarios overlook ways of knowing and being that, if given a chance, would offer counter-imaginaries and alternative possibilities of navigating the Anthropocene crisis.

Doomsday imaginaries in the museum of the Anthropocene

Thousands of miles away from the Yanomami lands, in Rio de Janeiro, the Museum of Tomorrow announces the arrival of the Anthropocene. Established in the run-up to the 2016 Olympic Games, the museum aims to engage the broader public in reflecting on the destiny of human and planetary life. Housed in an impressive building characterised by its organic architecture, the museum's main exhibition is organised into several rooms of a minimalist design. With few directives and scarce staff, the structure of the space inspires self-direction and reflection throughout. In the first room, a dark space with a dome ceiling, a projection map shows the genealogies of time, matter, and biological life. The video recreates a cosmological trajectory starting with the Big Bang and ending with the evolution of life on the planet. The final part of the video hints at the advent of human intellect and the story of Western techno-energetic progress, a topic that is further explored in the Anthropocene gallery.

After this vivid projection of the cosmos, visitors are guided into an area dedicated to exploring the Earth as an organically interconnected system.

In a dark square room, multiple wildlife photographs and projections draw the audience's attention to the ecological entanglements that compose the intricate web of life. Here, interactive exhibits explore genetic and ecological diversity, as well as the cascading consequences of mass extinction caused by human-induced factors.

Later in the exhibition, a radically different space introduces the Anthropocene. In a performative reflection of its technocentric politics, the Anthropocene manifests in a circular installation of six ten-metre-high digital totems. On the screens, a looping video features contemporary scenes of environmental momentum: wild tornados, extensive floods, coral bleaching, and industrial smoke covering the sky. Intermingled with the cataclysmic footage, data charts show the scientific metrics that designate the Anthropocene as a new unit in geological time: rising global temperatures, mounting CO_2 levels in the atmosphere, high deforestation rates, and escalating mass extinction, among others.

Committed to providing the scientific evidence for the Anthropocene epoch, the museum draws on the thesis of the Great Acceleration, that is, the post-Second World War era of continual economic growth and technological development (Steffen et al., 2015). Showing the phenomenal growth of the human enterprise in the form of economic activity, consumption, and resource use, the period of the Great Acceleration has become the widely accepted date for the onset of the Anthropocene. Not only has this thesis been conceptually welcomed; its hockey stick graphs have become the Anthropocene's visual symbol (Steffen et al., 2015).[3]

Few people would object to the statement that the Museum of Tomorrow constitutes an impeccable work of curatorial assemblage and exhibiting praxis. Not only is its digital exhibition impressive, but the statistics and quantitative figures displayed in the Anthropocene gallery are updated on a regular basis. As such, the planning of the gallery is almost exclusively oriented towards incorporating the scientific bases of the geological Anthropocene, as though this scientific "correctness" alone is sufficiently persuasive. Conversely, building environmental awareness is not always straightforward, as the reception of seemingly universal knowledge tends to be shaped by the values and local circumstances of those to whom it is being communicated. In referring to the science of climate change, sociologist Bruno Latour (2018) argues that the problem of denialism cannot be solved by producing ever more data and scientific evidence. On the contrary, climate deniers hold onto their opinions more strongly when faced with facts that contradict them. In this respect, science-based evidence and geophysical models alone may not be convincing enough to produce critical engagement with the Anthropocene, let alone stimulate collective socio-ecological transformation. As Isabelle Stengers (2017, pp. 385–386) puts it, "climatologists may well sound the alarm, but their models do not provide the answer to be given to this alarm."

Alongside the models and graphics, the Anthropocene gallery features a juxtaposition of images and music that produce an ambience that is equally captivating and distressing. Dramatic and awe-inspiring, the scenes of a ruined planet prompt mixed reactions of aversion and attraction. This experience resonates with what Kenneth Little (1991 as cited in Evans and Ryan, 2019, para. 4) dubbed as "ruin porn": "imagery that highlights a tendency for voyeurism and an erotic pleasure taken in the act of looking at objectified nature."

Almost exclusively, the museum's representation of planetary ruin consists of drone perspectives and bird's-eye views. Generic human crowds, urban sprawl, extensive grasslands, and wide-ranging landfills are depicted through this spatial perspective. However well-intentioned, these aerial vistas depersonalise the crisis and reinforce the perception of an emergent scientific mastery over nature (Demos, 2018). Invested in large-scale aerial imagery, the Anthropocene gallery reveals a lack of reflection about the implications of this type of technological mediation within specific political and economic frameworks. As Krungen (2013 as cited in Demos, 2017, pp. 16–17) notes, "[mapping imagery] comprises a visual system delivered and constituted by the post-Cold War and largely Western-based military-state-corporate apparatus." This type of imagery "offers an innocent-seeming picture that is, in fact, a 'techno-scientific, militarised, 'objective' image'" (Krungen, 2013 as cited in Demos, 2017, pp. 16–17).

Indeed, technological visualisation is never neutral, particularly in the geography of the Global South, where its use has been historically implicated in exploitative relations of natural resource extraction. By uncritically reproducing bird's-eye views and satellite imagery, the museum ignores the role of technology in mapping and commodifying the landscape. As such, the gallery reproduces a "colonial view," that is, a dominant viewpoint that advances territorial relations implied in the extractive practices of dispossession within capitalist accumulation. As Macarena Gómez-Barris (2017) argues, technological visualisation reinforces globalised modernity's separation from the agency of situated bodies. These visual representations occlude the normative ways in which we see and universalise the planetary, detaching it from the domain of lived experience. In the case of the museum, this perspective obscures the everyday practices and embodied vernacular knowledges existing in the Indigenous and Afro-Indigenous spaces of South America. This exacerbates the undervaluing of the experiences of local social and ecological movements, as well as questions of environmental justice in Indigenous and rural lands.

In sum, the Museum of Tomorrow positions modern science and technological development as the answer to our adaptation to climate change and other damaging effects of the Anthropocene. Embracing the disembodied model of Western science, the Museum of Tomorrow overlooks the fact that Indigenous lifeforms and local epistemes may have something to offer. Moreover, the museum remains oblivious to a long history of Indigenous environmental struggles and rural mobilisations for climate justice. In its galleries, there is no mention of Indigenous participation in the United Nations' Conference of the Parties (see Callison, 2015; Watt-Cloutier, 2015), or the courageous resistance of Indigenous groups to the imposition of mining and drilling projects in their ancestral lands (see Athayde, 2014; Estes, 2019; Gómez-Barris, 2017). In view of these absences, I contend that this exclusion represents not only a curatorial omission, but a political position. In excluding Indigenous voices from its representation of the Anthropocene, the museum replicates a form of epistemic erasure, that is, a hierarchical formation that positions scientific knowledge as the only valid perspective, dismissing ancestral or traditional knowledges as anecdotal, mythological, or folkloric (Castro-Gómez, 2007; Quijano, 2000).

The end as a beginning: worlding in the Anthropocene

Seen as a moment of uncertainty and conflict, the Anthropocene has often been presented through apocalyptic narratives and cultural imaginaries of civilisation collapse (Cook and Balayannis, 2015; Ginn, 2015; Swyngedouw, 2013). Amidst these apocalyptic readings, questions have been raised about the type of futures that the Anthropocene presents, and the kind of narratives it privileges. Among those interrogations, the voices of Indigenous peoples have been of most significance, as they have drawn connections between the process of colonisation and the idea of the apocalypse. Indeed, the colonisation of the Americas entailed not only the plunder of natural resources but the eradication of local peoples' forms of self-determination over their bodies and lands. As Viveiros de Castro and Danowski (2018, p. 192) put it, "losses of world, demographic catastrophes, and ends of history, means simply this: for the native people of the Americas, the end of the world already happened – five centuries ago. To be exact, it began on October 12, 1492."

Attempts to relate European colonisation to the chronologies of the Anthropocene have resulted in a proliferation of debates and disagreements. For instance, in a motion to foreground the radical disruption of ecologies that followed the arrival of Europeans to the Americas, geographers Simon Lewis and Mark Maslin (2015) have proposed the year 1610 as the start of the Anthropocene. Dubbed the Orbis Spike, this date provides the stratigraphic evidence for the ecological and sociological revolution of the "Columbian encounter." The geological mark, observed as a decline of CO_2 levels in the Antarctic ice cores, was the result of a mass reduction in farming due to a decrease in population and the consequent carbon uptake by vegetation and soils. Whilst presenting reliable evidence for a chronological marker, the Orbis Spike thesis has been criticised for being devoid of any political intention (Pulido, 2018; Yusoff, 2019). As Yusoff (2019) writes, through the euphemism of human exchange, 1610 disguises colonial relations of domination and submission, masking the mutilation of land and bodies, and the radical interruption of spiritual relations that existed before the invasion. Put another way, "in the language of exchange, it might be assumed that something was given rather than just taken" (Yusoff, 2019, p. 1).

In a similar vein, Anishinaabe scholar Kyle Whyte (2018) has drawn attention to the links between colonial violence and the Anthropocene, arguing that Indigenous peoples live in "dystopian or post-apocalyptic times." Thus, he notes that whilst academics, journalists, and artists continue to conjure apocalyptic portrayals of the future, the very same hardships they name – species loss, ecosystem degradation, economic crash, and social upheaval – have already been endured by Indigenous people due to different forms of colonialism. Indeed, the arrival of European settlers to the Americas inaugurated a new historical moment characterised by the plunder of ecosystems and entire nations alike: "an apocalyptic death-world of rape, poaching, trespass, theft, and disease" (Estes, 2019, p. 8). In this scenario, wildlife extinction was not lived as the simple elimination of species, but as the death of complex mutual relations cultivated with plants, animals, and the land. As experienced by the Oceti Sakowin First Nation,[4] the Buffalo Nation (Pte Oyate) was erased from the North American plainlands at the hands of fur and meat traders. For Oceti Sakowin and other First Nations, the buffalo population was a sister nation whose presence determined much of human existence in a network of kinship and mutual respect. Against this backdrop, the loss of the buffalos meant the permanent end of a world of shared history and care (Estes, 2019). As the Crow nation tribe's chief Plenty Coup put it, "when the buffalos went away, the hearts of my people fell to the ground, and they could not lift them up again. After this, nothing happened" (Linderman, 1930 as cited in Cronon, 1992, p. 1366).

Colonial violence faced by Indigenous people in South America was not so different from the brutal subjection of First Nations in the northern part of the continent. For Amerindian tribes, historical occupation of their territories has been the main source of decimation and death since European colonisation began in the sixteenth century. In the Amazon rainforest, for instance, the arrival of Portuguese colonisers and missionaries was lived as the advent of slavery, disease, pollution, and death (Eichen, 2020; Starling and Schwarcz, 2018). Alas, those deadly encounters were not limited to the first imperial expeditions, but extended well into the formation of modern-day Brazil. As Davi Kopenawa and Bruce Albert (1999) recount, Kopenawa's village was almost exterminated by the flu and measles, both introduced by public servants and evangelical missionaries in the 1950s and 1960s.

To this day, the shockwaves of colonial violence continue to shape the vast territories of the Amazon, perpetuating ecocidal and genocidal practices by means of land grabbing, gold and mineral prospection, and extensive monoculture. Within this scenario, it could be said that the advent of the Anthropocene has exacerbated the confrontation between divergent cosmological projects and their corresponding world-making practices.

The opposition between ontological projects resonates with Bruno Latour's description of the Anthropocene as a war between two fronts: Humans and the Earthbound. In his renowned Gifford Lectures, Latour (2017) depicts two sides of the bellicose contraposition. On one side, Humans or "moderns" are those who believe it will be possible to go on living in the unified, indifferent nature of the Holocene. On the other, Earthlings or "non-moderns" are those who are ontologically and politically bound to the Earth's cause. In this configuration, Humans have denied the Earth by defining themselves as technologically exempt from nature's trials. Conversely, Earthbound – or Terrans, an alternative translation of the French word *terriens* (Danowski and Viveiros de Castro, 2016) – are those who do not conceive of themselves as separate from the material existence of the Earth, and they include Indigenous societies. Put differently, Humans see themselves as detached from the Earth, lacking a sense of self-localisation that allows them to remain indifferent to the consequences of their actions. The Earthbound, in contrast, can be termed sensitive and responsible; they belong to a territory and are eager to deal with the world's current state of disarray.

Latour's proposition of a war between two worlds is a useful point of departure for exploring the different eco-political projects that operate within the framework of the Anthropocene crisis. His idea of two fronts provides a way of examining how different epistemologies and worldviews engage with the widespread effects of climate change and environmental degradation. However, Latour's categorical dichotomy appears limiting as he overlooks the spaces of confluence between different world-making practices. Here, I follow Blaser and de la Cadena (2018) when they state that it is hard to trace the lines between one camp and the other, as most humans are victims and culprits at once. Thus, whilst there is no definitive answer to Latour's question of who the Humans and the Earthbound are, it becomes important not to lose sight of the differences between two types of assemblages: those that are thoroughly invested in the practices that generate the Anthropocene, and those that are more or less forcefully dragged along by those practices. Following this line, I offer a comparison between the Museum of Tomorrow and *The Falling Sky*, in order to analyse the political practices and figurations of the future advanced by the two cultural expressions.

As a growing number of scholars have pointed out (e.g. Hickey, 2019; Ngo, 2019; Rifkin, 2017; Watts, 2013), different conceptions of the world imply different understandings of time. For instance, at the Museum of Tomorrow a recurrent temporal narrative occurs as deep time,[5] and a series of geophysical transformations are chronologically represented throughout the galleries of the museum. According to this narrative, at the end of this timescale, during the Anthropocene, the human species goes from being a simple biological agent to a geological force, signalling a singular bookend to a unitary planetary story. By contrast, *The Falling Sky* presents a more modest narrative of humanity's place in time-space configurations. Situated in mythological time, the story of the Yanomami is not one of human's dominance but one of interconnectedness between people, animals, plants, and spirits. Unlike the Anthropocene's "self-indulgent and self-fulfilling myths of the apocalypse" (Haraway, 2016, p. 39), Kopenawa's words offer a way of thinking about the urgency of these times with the conviction that it is still possible to re-pattern the webs and processes that interlace life. In this view, Yanomami cosmologies contest Euro-Western thought precisely because they differ from its aim of positioning the human as the pinnacle of the natural world. As Viveiros de Castro's (2012) concept of perspectivism illustrates, many Amerindian cosmologies are based on the postulate that, in ancestral times, the multiple entities of the cosmos shared a generic human condition and were thus able to communicate with each other. Myths often describe how, at some point, this condition suffered severe disruption, which resulted in the transformation of the numerous types of humans that existed into the diverse present-day species of animals, plants, and other beings. Whilst in ancestral times all beings were perceived as human and non-human at the same time, myths tell how they permanently adopted the animal (or other) bodies they have today (Vanzolini and Cesarino, 2014).

However, contemporary praxes like shamanism and hunting reveal that the previous human condition was not entirely overcome, since animals, objects, and spirits can still reveal an inner human form, usually associated with what is commonly translated as their "soul" or "double." External non-human appearance is thus often conceived as the skin or clothing that hides a human interior. This body or skin is responsible for determining a specific point of view. Thus, rather than an opposition between internal human essence and external non-human appearance, these cosmologies postulate a radical relationalism:

> while viewed by humans as animals, animals and other beings view themselves as humans and live in conditions similar to humans; that is, they have a social life similar to those who inhabit Amerindian villages. Humanity is thus the reflexive state of a subject to itself, while animality is the condition of the body regarded from an external point of view.
>
> (Vanzolini and Cesarino, 2014, para. 2)

Paradoxically, Amerindian perspectivism contains an anti-anthropocentric potential because, after all, "in a world where everything is human, being human is not that special" (Avelar, 2013, p. 255).

Overall, it could be said that shamanism in Yanomami culture is an eco-political praxis, in which the shaman escapes from the limits of a human perspective into the social worlds of other species; what Kopenawa calls "becoming other" (Kopenawa and Albert, 2013, p. 46). Shamans thus play the role of mediators, serving as interlocutors between the different societies of the forest.[6] Contrasting with most Anthropocene Western discourse, in which non-humans and even certain types of humans do not have a voice (Davis and Todd, 2017; Simpson, 2018; Todd, 2015; Yusoff, 2018), *The Falling Sky* is indeed a cosmo-ecological manifesto as its words bring forward the animacy and agency of non-human and inhuman entities (Arregui, 2019).

Regarding the concept of time, *The Falling Sky* is equally divergent, as it envisages Yanomami existence in a temporal horizon historically negated for Indigenous people: the future. In this view, Yanomami storytelling can be described as a future-oriented practice, or a form of futurity. As Baldwin notes (2012 in Hickey, 2019, p. 166), futurity refers to how "the future is rendered knowable through specific practices (i.e. calculation, imagination, and performance) and, in turn, intervenes in the present through three anticipatory logics (i.e. pre-caution, pre-emption and preparedness)." Yanomami futurity, then, comprises the structures and narratives that support and imagine a future for Yanomami people. In this sense, Indigenous futurities pose a direct challenge to Euro-Western notions, which tend to locate colonised and racialised peoples as perpetually past to European cultures (Quijano, 2000). In sum, this Indigenous self-conception of time resists yet another form through which colonialism operates. As Helen Ngo (2019, p. 248) puts it, "in addition to the continued material, political, and legal injustices of colonialism, the processes of colonisation and racialisation have also differentially temporalised, colonised and racialised bodies, pegging them to the (closed) past."

Furthermore, in continuously revising their understandings of the world, Indigenous people resist colonial attempts to catalogue their stories as folklore, myths, or legends (Watts, 2013). As *The Falling Sky* illustrates, Yanomami shamans persistently develop different versions and adopt new elements over time, including those falling under environmental discourse. In recent decades, Amerindian peoples have increasingly adapted Western environmental rhetoric in their cosmologies, conceptual vocabularies, and existential projects. With this, they have translated their world-making practices into a new language whose political intent is unequivocal. Remarkably, their message has started to make itself heard among the citizens of privileged societies in the Global North (Viveiros de Castro and Danowski, 2018).

In sum, whilst particular social and political spaces – such as the Museum of Tomorrow – uphold discourses of the apocalypse that reinforce an end-of-time social paralysis, in other contexts – such as the Yanomami lands – conceptual understandings of the end of the world can be the source of forceful political mobilisation. Indeed, and as a growing chorus of scholars and intellectuals have pointed out (e.g. Baldwin, 2017; Hickey, 2019; Rifkin, 2017; Whyte, 2018), dominant imaginations and visualisations of the future tend to drastically omit certain worldviews and conceptions of time. Against this backdrop, foregrounding hitherto *othered* experiences and knowledges can help us to identify the kinds of obstacles that must be overcome if we are to attempt to move towards a future that is more genuinely shared.

Conclusions

This chapter has sought to highlight Indigenous cosmologies as a form of eco-political resistance in a time of climate and ecological emergency. In particular, I have drawn attention to the ways in which Amerindian cosmologies uphold embodied, situated, and contextualised conceptions of "the end of the world," in contrast to the more dominant apocalyptic rhetoric of the Anthropocene. To that end, I have presented a comparative approach to two different cultural productions: the Museum of Tomorrow in Rio de Janeiro and *The Falling Sky*, an anthropological account of Yanomami shamanic cosmology.

In looking at the museum's curatorial endeavour, I have discussed its visual representations of climate change and ecological decline. In this respect, I have argued that a Western-centric vision of a global catastrophe prevails in the gallery of the Anthropocene, naturalising a view of the ecological crisis which is dominated by doomsday scenarios as well as numeric graphs and geophysical data. By addressing the Anthropocene through the lens of a scientific rationale, I have argued that the museum has overlooked the memories and experiences of Indigenous communities, and their forms of traditional, cosmological, and embodied knowledge concerning climate and environmental change.

As a contrast to the museum's visualisations of the Anthropocene, I have engaged with *The Falling Sky* as a powerful, cosmo-ecological manifesto. That is, a text that evidences that Yanomami cosmology, rather than mere folklore or a set of legends, is an integral part of Amerindian political resistance. In this respect, Yanomami storytelling has long informed political mobilisation against the extractive and capitalist forces that lie behind the profound ecological damages being propagated across the planet.

Moreover, rather than signalling a one-time event culminating with human extinction, Kopenawa's shamanic tales speak of world-beginnings and world-endings that are immersed in the process of planetary self-organisation (Szerszynski, 2019). Hence, Yanomami cosmologies foreground fluidity and transformation, interlinking different entities that might otherwise be seen as wholly separate categories (i.e. nature/culture). In sum, unlike Western representations of the human as a singular dominant entity, lifeforms in the Yanomami forest "exist in and are defined by meshworks of reciprocity and generosity, where everything is alive, aware, and a potential interlocutor" (Szerszynski, 2019, p. 204).

Building on Kopenawa's writing as a theoretical understanding of the world through physical embodiment (Arregui, 2019), I have drawn attention to the multifaceted forms of violence faced by the Yanomami people, who are immersed in a geographical context that bears the brunt of the extractive capitalist agenda (Blaser and de la Cadena, 2018; Gómez-Barris, 2017; Pulido, 2018). In this respect, my intention has been to foreground the fact that whilst some privileged groups – mainly in Euro-Western and Westernised societies – continue to work with scholarly and cultural understandings of a universalised Anthropocene, less privileged groups continue to be the victims of the Anthropocene's most damaging impacts on a daily basis. At the time of writing, Yanomami people are facing the violent expansion of extractive industries into their territory, not only via the intrusion of illegal logging and mineral prospection but also through the arrival of pollution, diseases, and new epidemics.[7] What is more, Yanomami and other Indigenous groups across Brazil continue to resist the necropolitical moves of far-right populist Jair Bolsonaro, whose open aversion towards Indigenous peoples endorses societal racism and justifies the illegal invasions of their lands.

Against this backdrop, and alongside many other voices from the South American continent (e.g. Blaser and de la Cadena, 2018; Krenak, 2019; Tironi, 2019; Viveiros de Castro and Danowski, 2018), I have argued that the Anthropocene is a time for the repositioning of agencies and politics. Devoting particular attention to the geographies of Brazil, I have claimed that Anthropocene politics must necessarily involve Indigenous agency, which has long been neglected, oppressed, effaced, or fixed in some essentialised past.

For that matter, non-Indigenous scholars should begin by decolonising our own perspectives on the Anthropocene. This endeavour should not be taken lightly, as it entails a serious response to questions of Indigenous and First Nations justice. Thus, it entails not only the need to emphasise colonialism's role in environmental change, but a need also to recognise the multiple and extended forms in which colonial logics continue to operate in our academic, Western-centric spaces. As Simpson (2018) notes, as long as the concept of the Anthropocene continues to privilege Western scientific knowledge, it will continue to exclude and silence other ways of knowing and being.

Overall, in examining the ways in which Amerindian peoples are actively resisting the "end of the(ir) world," I have argued that we – non-Indigenous people – should commit to breaking away from linear imaginations of the apocalypse. In so doing, we can work to challenge our long-standing notions of time, and to welcome circular, spiralling, or rhizomatic conceptions of time. As the stories of Yanomami people teach us, history is not what happens in a continuous timeline, but what occurs between the cyclical falling and rising of the sky. Perhaps only when we understand this, will we stop fearing the end of the world, and start appreciating the idea of worlding in its truest sense.

Notes

1 I acknowledge that the term Indigenous has been subject to numerous controversies and contestations. Whilst recognising that it would be unwise to generalise its use as an adjective, I employ the term for pragmatic reasons as a means of naming those communities holding an ancestral culture that responds to their own processes of historical construction (Macas, 2010).

2 For Kopenawa, white people (*brancos* in Portuguese or *näpë* in Yanomami) are all non-Indigenous people, whether Brazilian or foreign. It is important to note that the term *näpë* was originally used to name the relational and changeable condition of "enemy." Later on, it became a referent for white people, that is, the members of any national society aiming to destroy the political autonomy of the native people in question (Viveiros de Castro, 2010).

3 Even the logo of the Anthropocene Working Group, the interdisciplinary research group dedicated to examining the Anthropocene as a geological time unit is a visual reference to the Great Acceleration.
4 The Great Sioux Nation or the Nation of the Seven Council Fires wrongly named as Sioux by settlers (Estes, 2019).
5 Deep time is the scale through which gradual geological changes are often understood.
6 From an Amerindian perspective, the forest, or the environment, can be described as "a society of societies" (Viveiros de Castro and Danowski, 2018, p. 180).
7 For decades, Yanomami people have endured the introduction of measles, flu, tuberculosis, malaria, and other infectious diseases into their lands. Since March 2020, they have fought the spread of COVID-19 in their villages (Albert, 2020).

References

Albert, B. (2010) Postcriptum. Quando eu é um outro (e vice-versa). In D. Kopenawa, & B. Albert (Eds.), *A queda do céu. Palavras de um xamã yanomami* (Trans. B. Perrone-Moisés) (pp. 512–549). São Paulo: Companhia das Letras.

Albert, B. (2020) COVID-19: Lessons from the Yanomami. *The New York Times [Online]*. www.nytimes.com/2020/04/27/opinion/yanomami-covid-brazil.html

Arregui, A. (2019) This mess is a "world"! Environmental diplomats in the mud of Anthropology. In R. Bold (Ed.), *Indigenous Perceptions of the End of the World.* (pp. 183–202). Cham: Palgrave MacMillan.

Athayde, S. (2014) Introduction: Indigenous peoples, dams and resistance. *Tipiti: Journal of the Society for the Anthropology of Lowland South America, 12*(2), 80–92.

Avelar, I. (2013) Amerindian perspectivism and non-human rights. *Alter/Nativas, 1*, 1–21.

Baldwin, A. (2017) Postcolonial futures: Climate, race, and the yet-to-come. *ISLE: Interdisciplinary Studies in Literature and Environment, 24*(2), 292–305.

Blaser, M., & de la Cadena, M. (2018) Pluriverse. In M. Blaser & M. de la Cadena (Eds.), *A World of Many Worlds* (pp. 1–22). Durham: Duke University Press.

Callison, C. (2015) Beyond COP21. Collaborating with indigenous people to understand climate change and the Arctic. In N. Aikens, B. Caka, D. Franssen, N. Haq, S. Lopez, N. Paynter, N. Petrešin-Bachelez, C. Prieto del Campo, S. ten Thije & A. Železnik (Eds.) *Ecologising Museums* (pp. 15–23). L'internationale Books. https://d2tv32fgpo1xal.cloudfront.net/files/04-ecologisingmuseums.pdf

Castro-Gómez, S. (2007) Decolonizar la universidad. La hybris del punto cero y el diálogo de saberes. In S. Castro-Gomez & R. Grosfoguel (Eds.), *El giro decolonial.* (pp. 79–91). Bogotá: Siglo del Hombre.

Chagnon, N. A. (1988) Life histories, blood revenge, and warfare in a tribal population. *Science, 239*, 985–992.

Cook, B. R., & Balayannis, A. (2015) Co-producing (a fearful) Anthropocene. *Geographical Research, 53*(3), 270–279.

Cronon, W. (1992) A place for stories: Nature, history, and narrative. *Journal of American History, 78*(4), 1347–1376.

Danowski, D., & Viveiros de Castro, E. (2016) *The Ends of the World.* Cambridge and Malden: Polity Press.

Davis, H., & Todd, Z. (2017) On the Importance of a date, or decolonizing the Anthropocene. *ACME: An International E-Journal for Critical Geographies, 16*(4), 761–780.

Demos, T. (2017). *Against the Anthropocene: Visual Culture and Environment Today.* Berlin: Sternberg Press.

Demos, T. (2018) To save a world: Geoengineering, conflictual futurisms, and the unthinkable. *E-Flux, 94.* www.e-flux.com/journal/94/221148/to-save-a-world-geoengineering-conflictual-futurisms-and-the-unthinkable/

Eichen, J. R. (2020) Cheapness and (labor-) power: The role of early modern Brazilian sugar plantations in the racializing Capitalocene. *Environment and Planning D: Society and Space, 38*(1), 35–52.

Estes, N. (2019) *Our History Is the Future: Standing Rock Versus the Dakota Access Pipeline, and the Long Tradition of Indigenous Resistance.* London and New York: Verso.

Evans, M., & Ryan, N. (2019) (De)compositions: A review of Anthropocene. *Visual and New Media Review, Fieldsights.* https://culanth.org/fieldsights/decompositions-a-review-of-anthropocene

Ginn, F. (2015) When horses won't eat: Apocalypse and the Anthropocene. *Annals of the Association of American Geographers, 105*(2), 351–359.

Gómez-Barris, M. (2017) *The Extractive Zone: Social Ecologies and Decolonial Perspectives.* Durham: Duke University Press.

Haraway, D. (2016) Staying with the trouble Anthropocene, Capitalocene, Chthulucene. In J. Moore (Ed.), *Anthropocene or Capitalocene? Nature, History, and the Crisis of Capitalism* (pp. 34–76). Oakland: PM Press.

Hickey, A. (2019) Rupturing settler time: Visual culture and geographies of indigenous futurity. *World Art, 9*(2), 163–180.

Kopenawa, D., & Albert, B. (1999) Descobrindo os brancos. In C. Inada (Ed.), *A Outra Margem de Occidente* (pp. 15–22). São Paulo: Companhia das Letras.

Kopenawa, D., & Albert, B. (2013) *The Falling Sky* (Trans. Elliott, N & Dundy, A.). Cambridge: Harvard University Press.

Krenak, A. (2019) *Ideias para adiar o fim do mundo.* São Paulo: Companhia das Letras.

Latour, B. (2017) *Facing Gaia.* Cambridge and Medford: Polity Press.

Latour, B. (2018) *Down to Earth: Politics in the New Climatic Regime.* Cambridge and Medford: Polity Press.

Lewis, S., & Maslin, M. (2015) Defining the Anthropocene. *Nature, 519,* 171–180.

Macas, L. (2010) Abya Yala and the decolonization of democracy, knowledge, education and the State. In L. Meyer & B. Maldonado (Eds.), *New World of Indigenous Resistance.* (pp. 239–250). San Francisco: City Lights Books.

Moore, J. (2016) Introduction. In J. Moore, D. Haraway, D. Hartley, C. Parenti, & J. McBrien (Eds.), *Anthropocene or Capitalocene?: Nature, History, and the Crisis of Capitalism* (pp. 1–11). Oakland: PM Press.

Morrison, K. D. (2015) Provincializing the Anthropocene. *Seminar, 673,* 75–80.

Ngo, H. (2019) 'Get over it'? Racialised temporalities and bodily orientations in time. *Journal of Intercultural Studies, 40*(2), 239–253.

Pulido, L. (2018) Racism and the Anthropocene. In G. Mitman, M. Armiero, & R. Emmett (Eds.), *Future Remains. A Cabinet of Curiosities for the Anthropocene* (pp. 116–128). Chicago: The University of Chicago Press.

Quijano, A. (2000). Coloniality of power and eurocentrism in Latin America. *International Sociology, 15*(2), 215–232.

Rifkin, M. (2017) *Beyond Settler Time: Temporal Sovereignty and Indigenous Self-determination.* Durham: Duke University Press.

Simpson, M. (2018) The Anthropocene as colonial discourse. *Environment and Planning D: Society and Space, 0*(0), 1–19.

Starling, H. M., & Schwarcz, L. M. (2018) *Brazil: A Biography* (anonymous, Trans.). New York: Farrar, Strauss and Giroux.

Steffen, W., Broadgate, W., Deutsch, L., Gaffney, O., & Ludwig, C. (2015) The trajectory of the Anthropocene: The great acceleration. *The Anthropocene Review*, *2*(1), 81–98.

Stengers, I. (2017) Autonomy and the Intrusion of Gaia. *South Atlantic Quarterly*, *116*(2), 381–400.

Swyngedouw, E. (2013) Apocalypse now! Fear and doomsday pleasures. *Capitalism Nature Socialism*, *24*(1), 9–18.

Szerszynski, B. (2017) Gods of the Anthropocene: Geo-spiritual formations in the Earth's new epoch. *Theory, Culture & Society*, *34*(2–3), 253–275.

Szerszynski, B. (2019) Epilogue: Indigenous worlds and planetary futures. In R. Bold (Ed.), *Indigenous Perceptions of the End of the World* (pp. 203–209). Cham: Palgrave MacMillan.

Tironi, M. (2019) An introduction to "The Anthropocene in Chile: Toward a new pact of coexistence". *Environmental Humanities*, *11*(2), 465–466.

Todd, Z. (2015) Indigenizing the Anthropocene. In E. Turpin & H. Davis (Eds.), *Art in the Anthropocene: Encounters among Aesthetics, Politics, Environments and Epistemologies* (pp. 241–254). London: Open Humanities Press.

Vanzolini, M., & Cesarino, P. (2014) Perspectivism. In *Oxford Bibliographies*. Oxford University Press. www.oxfordbibliographies.com/view/document/obo-9780199766567/obo-9780199766567-0083.xml

Viveiros de Castro, E. (2009) Intensive filiation and demonic alliance. In C. Bruun Jensen & K. Rödje (Eds.), *Deleuzian Intersections: Science, Technology, Anthropology* (pp. 219–253). Oxford and New York: Berghahn Books.

Viveiros de Castro, E. (2010) Prefácio—O recado da mata. In D. Kopenawa & B. Albert, *A queda do céu. Palavras de um xamã yanomami* (pp. 11–41). São Paulo: Companhia das Letras.

Viveiros de Castro, E. (2012) *Cosmological Perspectivism in Amazonia and Elsewhere.* HAU Masterclass Series Vol. I. Manchester: HAU Journal of Ethnographic Theory.

Viveiros de Castro, E. (2019) On models and examples: Engineers and bricoleurs in the Anthropocene. *Current Anthropology*, *60*(S20), S296–S308.

Viveiros de Castro, E., & Danowski, D. (2018) Humans and Terrans in the Gaia War. In M. de la Cadena & M. Blaser (Eds.), *A World of Many Worlds* (pp. 172–203). Durham: Duke University Press.

Watt-Cloutier, S. (2015) *The Right to Be Cold: One Woman's Story of Protecting Her Culture, the Arctic and the Whole Planet.* Toronto: Penguin Canada.

Watts, V. (2013) Indigenous place-thought and agency amongst humans and non-humans (First Woman and Sky Woman go on a European world tour!). *Decolonization: Indigeneity, Education & Society*, *2*(1), 20–34.

Whyte, K. (2017) Indigenous climate change studies: Indigenizing futures, decolonizing the Anthropocene. *English Language Notes*, *55*(1), 153–162.

Whyte, K. (2018) Indigenous science (fiction) for the Anthropocene: Ancestral dystopias and fantasies of climate change crises. *Environment and Planning E: Nature and Space*, *1*(1–2), 224–242.

Yusoff, K. (2018) *A Billion Black Anthropocenes or None.* Minneapolis: University of Minnesota Press.

Yusoff, K. (2019) White utopia/black inferno: Life on a geologic spike. *E-Flux*, *97*. www.e-flux.com/journal/97/252226/white-utopia-black-inferno-life-on-a-geologic-spike/

4 Apocalypse repeated

The absence of the indigenous subject in George Turner's *The Sea and Summer* (1987)

Charlotte Lancaster

George Turner's 1987 novel *The Sea and Summer* begins a thousand years in the future with "The Autumn People", a community living in the highest reaches of the Dandenong Ranges above the drowned "old city" of Melbourne. Floating above the "old riverbed" of the Yarra, the pilot of a boat employed for tours of the "sunken city" surveys its "drowned ruins", which "had reached its swollen maximum of population and desperation a thousand years before" (3). Looking back at the cultural history of the "long summer" of the twenty-first century, the pilot remarks that "they and those like them ruined the world for all those who came after [...] they *denied* history" (7). *The Sea and Summer* presents a world defined by overwhelming environmental and sociopolitical breakdown. By the middle of the twenty-first century the global population has exceeded the ten billion mark, a significant rise in sea levels has resulted in the inundation of the world's coastal regions and the global economy has collapsed due to expensive geoengineering projects that have failed to alleviate the crisis. Australia struggles with mass unemployment, social schism and draconian governmental control, while the sea engulfs Melbourne's bayside suburbs, slowly submerging the high-rise tower blocks where the "Swill", the city's poor, have been consigned. Beyond the submerged zone of the "Swill" live the affluent "Sweet" communities, safe from the floodwaters on higher ground.

Turner achieved early accolades for *The Sea and Summer*, winning both the Commonwealth Writers' Prize Best Book Award for the South East Asia and South Pacific Region and the Arthur C. Clarke Award for best science fiction (SF) novel published in Britain in 1988. Yet, the book received marginal critical attention until its recent reissue in 2013 as the first Australian title to be included in Victor Gollancz's list of "SF Masterworks". It has enjoyed a small upsurge in interest in recent years, perhaps due to its seemingly prophetic vision of climate crisis in the twenty-first century. As one of the earliest novels to depict climate crisis as a consequence of unfettered anthropogenic warming, *The Sea and Summer* offers a vision of environmental collapse that predates other significant post-apocalyptic representations of climate crisis in Australia, such as Jane Rawson's *Wrong Turn at the Office of Unmade Lists* (2013) and James Bradley's *Clade* (2015). The proliferation

DOI: 10.4324/9781003128854-5

of these environmental apocalyptic fictions serves to show that apocalypse remains a potent trope in the Australian environmental imagination. Yet, the genre also remains highly problematic with respect to climate crisis, not least because the apocalyptic imaginary, as Bruno Seraphin (2016) has observed, "speciously universalises a set of environmental anxieties that is in fact particular to white settler society" (144). With this in mind, I turn to Turner's depiction of environmental apocalypse to question the notion of dispossession and "*denied* history" expressed in the novel (7). It appears on a closer reading of *The Sea and Summer* that an imaginative gap exists between the seeming forgetfulness of the violence and illegitimacy of white settlement and the violence of the social caste system depicted.

Indeed, the community in the Swill Enclaves suffer repeated apocalypses due to their imaginative and physical marginalisation from the state. In this way, the narrative could be seen to echo the forms of colonial violence endured by Australia's Aboriginal Indigenous communities for previous centuries, such as economic crash, environmental collapse, displacement and cultural disintegration. However, indigenous voices remain startlingly absent from the novel. My reading of *The Sea and Summer* thus aims to foreground the veiled histories of colonial trauma denied by the novel's ahistorical presentation of environmental apocalypse. In order to fully sketch out my argument, I first give a brief overview of current scholarly discussion around the literary genealogy of environmental apocalypse and its generic conventions, before turning to the novel.

Apocalyptic rhetoric has long been thought of in ecocritical theory as a primary, and perhaps necessary, element in environmental discourse, "capable of galvanising activists, converting the undecided and ultimately, perhaps, of influencing government and commercial policy" (Garrard, 2004, p. 104). This is due to the meaning of apocalypse as a form of revelation or disclosure, from the Greek *apokálypsis* (uncovering) and *apokalyptein* (to uncover). Portrayals of environmental apocalypse, therefore, can be read not as end of the world narratives, but as revelatory narratives that disclose the potentially catastrophic consequences of our current trajectory to transform the collective mindset.

Despite this, Astrid Bracke (2018) notes that "the popular idea of apocalypse as "the end of the world" might stand in the way of productive, rather than defeatist, engagements with climate crisis" (25). Environmental apocalyptic fictions may offer the reader a dystopic vision of the future and mourn the end of the world as we know it. However, they do so in large part without acknowledging the histories of imperialism and (settler) colonialism that, as Kathryn Yusoff (2018) emphasises, "have been ending worlds for as long as they have been in existence" (xiii). In the context of this selective perspectivism, indigenous scholar and activist Kyle P. Whyte (2018) has called for researchers in the environmental humanities and beyond to "discuss how [such] concepts of finality would be received by Indigenous persons who see their societies as already having endured one *or many more* apocalypses" (236). Whyte's criticism speaks to a growing body of scholarship that directly challenges the ways in which narratives of crises, dystopia and apocalypse often work to obscure ongoing oppression against Indigenous peoples and other groups.

Mabel Gergan, Sara Smith and Pavithra Vasudevan (2018), for example, explore how particular apocalyptic and catastrophic engagements with climate crisis deploy a "temporal sleight of hand" which works to "escape specific culpability (for instance in processes of colonialism, capitalism, or imperialism) and instead centre a universal human frailty that ends with triumph, a clear moral, and a clean slate" (2). While the environmental humanities scholar April Anson (2017) notes, "[w]hen environmental apocalypse narratives do not explicitly deal with the history of the genre, they are in danger of reinforcing the anthropocentric CO_2lonialism at the root of climate change's attendant apocalyptic realities" (7). A genealogical reading of the apocalypse genre, therefore, is necessary to establish the ways in which apocalyptic portrayals of environmental crisis often fail to grapple with the legacies of violent dispossession of indigenous land and ongoing extractive economies that resonate under the auspices of colonialism.

According to Roslyn Weaver (2011), the apocalypse genre "forms one important strand in the overall fabric of [Australian] cultural identity" (3). Yet, the historicity of the genre uncovers a narrative of indigenous erasure and land dispossession, stretching back to pre-colonial European conjecture about *terra australis incognita*, or the unknown southern land, as a place that paradoxically inspired both new world hope, or horror and dystopia (3). The "perpetual and persistent speculation" about the continent, Weaver (2007) notes, "established an apocalyptic map of Australia before colonists even experienced the land, and has resulted in a tradition of imagining the nation in apocalyptic terms ever since" (iii).

As Weaver's historiography shows, apocalyptic discourse not only directly correlates with a subjective colonial experience of Australia, but also facilitated the culturally destructive processes of colonial European advancement and absorption by representing the country as *terra nullius*, "a blank slate upon which one might inscribe any image" (2007, p. 52). Early European explorers of Australia thus inscribed the land with themes and images which came to influence the experiences of British colonisation and settlement in the eighteenth and nineteenth centuries. Of these, the leitmotif of dispossession forms a recurrent strand in the Australian environmental apocalyptic imaginary, betraying an ambivalence about belonging and possession that is endemic to Australian settler society.

Writing on the conventions of apocalypse films such as the *Mad Max* (1979) franchise and *Dead End Drive-In* (1986), Claire Corbett (2017) argues, "[t]he trauma of Indigenous dispossession is both personal and political. It continues to reverberate nationwide, a stark example of how the past is never past, present and future cannot be disentangled" (339). Corbett's analysis invokes psychoanalytic theory to examine the ways in which Australia's colonial history is both remembered and deliberately forgotten through the apocalyptic imaginary that is "dislocated from the past to the 'future' and thus in effect to a timeless, ever-present or ever-recurring time" (329). Corbett relates the experience of repetition compulsion, understood by Freud as a symptom of post-traumatic stress disorder. Repetition compulsion causes the sufferer "to repeat the repressed material as a contemporary experience instead of [...] remembering it as something belonging to the past" (2015, p. 161). In the context of the traumatic legacy and "deep forgetting" of Australia's dispossession of Indigenous peoples, Corbett views the Australian apocalyptic imaginary as enacting a simultaneous remembering and not remembering. "Some of these images" she writes, "seem to be like the 'flashbulb' memories associated with PTSD, intruding into awareness and vanishing or sometimes encoded within other story elements, leaving traces" (338). Corbett's invocation of memory and trace is interesting in relation to the ways Australia's history of colonial violence continues to repeat across its apocalyptic literature.

As I will show, the trace of indigenous presence and absence similarly haunts Australia's environmental apocalyptic imaginary in *The Sea and Summer*. While *The Sea and Summer* can be seen to have simply erased any notion of Australia's Indigenous history, its depiction of the state's imaginative and physical dispossession of the "Swill" class, I argue, can be read as simultaneously evoking Indigenous experience under the auspices of ongoing colonial (settler) violence. In the remainder of this chapter, I extend the critical analyses previously outlined to focus on how environmental apocalypse functions in the novel as a trace of colonialism and its legacies, nationalism and globalisation.

In his often-quoted study, *From Apocalypse to Way of Life* (2003), Fredrick Buell argues that by the 1980s, environmental collapse was not represented as a future apocalyptic disaster, as seen in environmental thought and writing from the 1960s and 1970s, but as a "way of life", or a "domestication of" or "dwelling within" environmental crisis (173, 205). *The Sea and Summer* can be seen to reflect these broader changes in depictions of environmental and climate crisis, where characters live in a world in which past prefigurations of unbridled carbon emissions have already been exceeded. Having "started a sequence which had to run its course in unbalancing the climate", the novel's characters dwell in the all-pervading motion of climatological and environmental change (14). While *The Sea and Summer*'s characters struggle with the daily realities of environmental collapse, they often distance climate crisis as an end-of-the-world future scenario. Seen through the eyes of Francis Conway, for example, environmental collapse exists as "a whining voice at the back of my mind", which "insisted that while the greedy ocean rose, year by year, the real catastrophe was yet to come. Behind that again was the cowardly whisper of humanity in all ages: "Please, not in my time" (180). Even as the novel's representation of climate change resonates with Buell's notion of a postmodern "domestication of environmental crisis", Francis Conway's doom-laden language shows how the popular appeal to apocalypse as the end of the world may distance or undermine the reality of a crisis that is already occurring. Francis's cognitive dissonance offers a reflection of the way the novel problematises the role of the apocalyptic narrative in interpreting environmental crisis. Climate crisis is depicted in *The Sea and Summer* through the breakdown of the environmental conditions that have fostered the growth and spread of consumer-driven capitalism. It focuses, for the most part, on the gradual deterioration of the urban cityscape, where the disintegration of capitalist infrastructures has produced a chaotic post-apocalyptic space, wrought by deep sociopolitical divisions.

The absence of cars, coffee, technology and even books comes to represent the collapsing state of consumer culture, and accordingly projects the "end of the world" as the end of capitalism. In place of consumer plenty and liberal freedoms, Turner depicts a world where all remaining resources are controlled by those with remaining capital in alliance with state power, while the rest of humanity struggles for mere survival. "Nine of every ten of Australia's population were Swill", one of the protagonists explains, "and many other countries were in worse case. Living familiarly with such knowledge, the horror of it passed us by; it was the normal condition of the world" (24). Yet, as many environmental scholars have argued, depicting the "normal condition" of the post-apocalyptic world after the death of capitalism as one without order, access to food, water and healthcare detaches the reader from the fact that these conditions describe the past and present realities of many people in the world. As Claire Colebrook (2017) notes in her investigation of Anthropocene culture: "What post-apocalyptic texts depict as the end of the world are counter-factual thought experiments of a humanity that has defined itself through the liberal privileges and favourable conditions that have drawn upon and depleted the worlds of others" (103). In this respect, *The Sea and Summer* exemplifies the highly problematic propensity for contemporary apocalyptic narratives to overlook the connection between the conjoined processes of capitalism, colonialism and imperialism with the experience of environmental collapse.

The Sea and Summer is comprised of two interchanging narratives. The first provides a framework for the main narrative and is told from the distant future perspective of the Autumn People. While this future has eliminated poverty and hunger from the world, it is by no means utopian. Indeed, the characters frequently discuss the effects of the impending global ice age, or "Long Winter", presumably due to accelerated effects of anthropogenic warming on the world's glacial and interglacial cycles. The second narrative, which forms the main body of the novel, features as a "Historical Reconstruction" of the development of the Greenhouse Culture through the years 2044–2061. Also entitled *The Sea and Summer*, it has been written by Professor Lena Wilson of the Autumn People, a historian specialising in the collapse of the Greenhouse Culture in Australia.

The story centres on the Conways' fall from grace in the Sweet colonies due to the father's job loss and subsequent suicide. Without an income, but with some savings, the Conways are relegated to the Fringe – a suburban area existing between the higher grounds of the Sweet and the drowning enclaves of the Swill. Alison employs the aid of Billy Kovaks, a small-time "tower-boss", to keep her family safe from the supposed violence of the Swill communities. The rest of the narrative continues to elaborate on the environment of the enclaves and the sociopolitical class divisions actively encouraged by the state in order to "preserve an economically manageable status quo" (157). The narrative concludes with the state-induced introduction of a virus to decrease the population that is smuggled into the enclaves via an addictive drug called "chewy". The main narrative is polyphonic, tracing the lives of the Conway family (Alison Conway and her two sons, Francis and Teddy), and their associates (Nola Parkes, Captain Nikopoulos and Arthur Derrick). The novel is written counter-chronologically, beginning and ending in 2061, but also weaving in and out of the 2040s and 2050s in time with the characters' memories.

In this respect, the novel takes a non-linear approach to time, suggesting that the linear approach offered by conventional narratives is insufficient to capture the vast spatiotemporal scales of environmental time. Furthermore, both the main narrative and the framework narrative conclude with open endings, illustrating environmental apocalypse as a recurrent and evolving phenomenon. In this way, Turner's narrative framework undermines the traditional notion of apocalypse as a catastrophe to be encountered in the future. By placing apocalyptic narratives in different temporal settings, *The Sea and Summer* places emphasis on history, not only as a written record of the past, but as a living force. In this way, the structure of the novel opens up questions about the role of narrative perspective in representations of climate crisis.

The novel frames the wilful deafness regarding public knowledge of the history of the Greenhouse Culture as a problem of narrative. In the framework narrative of the Autumn People, Lena intends her historical reconstruction of the Greenhouse era to function as a correction to "the public idea of what our ancestors were like. All they have is folklore and guesswork and idiotic popular plays that can't get so much as the clothing right" (16). Lena's concern about correcting the public idea or story of Australia's past is akin to the ways in which the dominant narrative of white settler history in Australia has repressed and silenced Indigenous perspectives. For example, the much-discussed recent *Dark Emu, Black Seeds: Agriculture or Accident* (2014), by Aboriginal author and historian Bruce Pascoe, highlights settler colonial inability and refusal to recognise Aboriginal land-use practices as farming, thus feeding into the "terra nullius" myth. Lena emphasises that her work is a "true story" because it is based on "records on tape and in data banks. There are descriptions, even pictures and police records providing detail" (16). While Lena highlights the socio-economic tensions in her historical reconstruction of Melbourne in the Greenhouse era, her overall response to the global climate crisis presents disturbance and risk as a constant for the majority, if not all, who appear equal in culpability as well as exposure. As such, Lena's apocalyptic narrative functions as a neoliberal discourse of adaptation, which, as Greg Bankoff (2019) explains, accepts disaster as an endemic condition of climate crisis while also obscuring "questions about the role of power and culture in society, and about whose environments and livelihoods are to be protected and why" (234).

Yet, while the text often challenges these connections, Lena's interjections highlight state oppression as an accidental, and even necessary, reaction to environmental collapse: "With the collapse of trade and all essential industry the Swill became a burden on the economy, easier and cheaper to support if it were concentrated into smaller areas" (106). Moreover, when Andra, attempting to grasp the social inequalities of the Greenhouse era, asks: "*How did this division arise? Why no revolution?*", Lena suggests that the "rise of the Tower Bosses" helped to run "small states within the State" and therefore allowed the Swill "a measure of contentment [...] by letting them run their own affairs" (106). In speaking for the Swill population, and speculating on their sociopolitical dispossession, Lena undermines the reliability of her narrative as a comprehensively accurate history. The Swill not only demonstrate an acceptance of their oppression in Lena's speculation, but are also shown to be complicit in such oppression, revealing a more ambiguous relationship to the processes of capitalist expansion and intensification. Lena's characters, therefore, appear as ventriloquial, dramatised embodiments of her own assumptions about the Greenhouse years as the meta-authorial voice of the text.

The history of the Greenhouse Culture is thus rendered suspect with its conflation of fact and fiction, suggesting that, despite best intentions, narratives are inherently unreliable. The instability of Lena's narrative is further compounded when Andra is forced to abandon his theatrical adaptation of the work due to his continued lack of comprehension of the Greenhouse years. "It is too easy to fall into the trap of seeing history in terms of human movement, as though all else is ancillary, as though *we* make history" he observes (361). It is clear that Andra's difficulty in imagining the history of environmental collapse is not so much due to his incapacity to imagine a temporality beyond the human, but more to do with how he imagines the "*we*" of humanity. Sylvia Wynter (2003) has argued that the conceptual category of the "human" often refers to a particular "ethnoclass (i.e., Western bourgeois)" that "overpresents itself as if it were the human itself" in narratives about the Anthropocene (260). As I have previously mentioned, critics have identified this as a problematic facet of the environmental apocalypse genre, which extends an idea of a homogenised humanity based on Western assumptions while deprivileging others. We can see this at work throughout Lena's conversations with Andra. "In the twentieth and twenty-first centuries the entire planet stood with its fingers plugging the dykes of its own creation" she tells Andra, "until the sea washed over their muddled status quo" (16). The humanity she describes of the mid-twenty-first century is analogous in regard to its impact on the planet to a species.

Narratives, the novel suggests, have the capacity to corrupt or be corrupted and therefore provide a false picture of the world. This is most explicitly addressed in relation to the Sweet's representation of the Swill throughout the novel as "dirty, criminal, ignorant", "animals" and, at times, "monsters" (150, 170). This dichotomy between Sweet and Swill is set up early on in the core narrative when Francis explains that "we kids were born to the knowledge that Sweet had jobs and income while Swill lived on State charity. Even servants could look down on Swill. Actually, very few Sweet kids of the day had ever seen a Swill person" (24). Not long after this we are told that the "Swill enclaves were not much mentioned in polite society; you knew of them and that was enough" (27). The novel succinctly highlights the double challenge the poor face as invisibility and amnesia.

While they might constitute the majority of the populace, the Swill have been forcibly marginalised in terms of visibility – placed on the away from the Sweet suburbs, where they are presumably out of sight and out of mind as the "high tide covers the ground levels of [their] tenements" (27). They also endure a form of public forgetting, not spoken of in "polite society" as Francis informs the reader. While sensationalist television programmes project an image of the Swill as violent criminals: "Swilldom" or "Swill horror country", Francis explains, "was also triv serial adventure country from which brave policeman or brave young scientists or brave and muscular foot-ballers rescued beautiful girls held captive for purposes not fully understood" (40). Andrew Baldwin (2017) has shown how the rapidly growing discourse on migration and climate crisis often consists of racist undertones. Within this, the image of the climate migrant, refugee or displaced person often bears striking resemblance to the figure of the monster.

Monsters, Baldwin writes, are "markers of and marked by disaster" and often embody the "lingering effects of disaster to admonish us to query the values that materialised the disaster" (296). Yet, it is often the case that monsters are not received as a signifier of the disaster, but come to be blamed for the occurrence of the disaster. Baldwin uses the example of the racist depictions of "black bodies criminalised in the wake of Hurricane Katrina [...] blamed for their "monstrous" condition, even while this perceived condition owes more to a collective failure to address the crisis of race in America" (296–297). Similarly, the figure of the Swill as the criminalised and dispossessed Other in *The Sea and Summer* could be read as an oblique reference to Australia's histories of racism.

When visiting the Swill tenements for the first time, Alison Conway is struck by the "monstrousness" of the tower blocks. "Each tenement tower was a blunt shaft assaulting the sky", she describes, "and around its base, like a dancer's skirt, spread 100 meters of concrete desert" (169). The representation of the tenements as an urban desert plays into historic Australian anxieties concerning the desert as a place of abjection, and therefore connotes the Swill enclaves and their residents as hostile. Weaver has observed that profoundly negative constructions of the Australian outback appear throughout colonial texts, and have been perpetuated in apocalyptic fictions to present the contrast between the power of the hostile landscape and the vulnerability of "civilised" humanity. Weaver (2007) notes

> [p]eople often refer to the centre of Australia – its 'heart' – as the red centre, and the negative version of this has been the 'dead centre', or the red heart, the 'never-never' and nothingness, a cursed place of punishment while hope for the future is away from the interior.
>
> (128)

Similarly, Roslynn D. Haynes (1998) has pointed out that early European descriptions of Australia as *terra nullius* characterise the Australian desert as a "hideous blank" (36). Depicted as an urban desert, the Swill tenements are likewise constructed as a "dead centre", into which the Swill have been physically and imaginatively "resettled" as "uninhabitants" of the nation state (Nixon, 2011, p. 153). The image of the Swill enclaves as a desert, an empty and hostile "dead centre" offers an example of indigenous absence and presence in the text.

As Haynes has observed, there is a long tradition of connecting Indigenous people with the outback and thus downgrading their status from inhabitant to "virtual uninhabitant" in associating Indigenous presence/absence with the vacant space of the desert. According to Haynes, the popular association between the desert and Indigenous habitation has several historical aspects:

> More than any other geographical region, the desert was associated with the Aborigines. This was the last stronghold to which they had been effectively driven by the expansion of settlement from the coastal fringes, and since it remained the area most inimical to Europeans it was readily ceded to them until it began to acquire commercial and strategic value for mining or nuclear testing. This association between Aborigines and the desert was not, however, a simple empirical matter; it resonated with political and ideological implications in a bitter preview of the land rights issue that has continued to bedevil Australian history.
>
> (34)

The forced conversion of Australian citizens from inhabitant into uninhabitant in *The Sea and Summer* can be seen as a trace of the imaginative expulsion of Indigenous peoples in this regard. The floodwaters of the rising tides in the text perform the constricting boundaries between the rich and the poor, compressing the urban space to a claustrophobic degree so that the poor communities become physically confined to the slowly drowning tower blocks. The recurring theme of sea level rise allows Turner to problematise spatial issues relating to imaginary borders between classes, cultures and nations to illustrate the country's very real struggles with discrimination, separatism and racism.

The movement of the Sweet away from the city towards Melbourne's suburbs is akin to the processes of "white flight", described by Carolyn Merchant (2003) as a recent example of how race has organised land use and power distribution in North America. Merchant observes how African Americans, who had once been associated with the "wilds" of Africa or the rural south, became symbols of decaying cities. "Sublime nature", Merchant writes, "was white and benign, available to white tourists; cities were portrayed as black and malign, the home of the unclean and the undesirable" (385). The idea of "white flight" is reflected in a similar way when Alison visits the Swill tenements for the first time. Recoiling at the sheer amount of people she sees on the tenement streets, Alison explains:

> Across the width of the roadway it stank of sheer dirtiness and sweat. [...] I saw without immediately understanding that in the packed desert of flesh a sort of territoriality rules. On the featureless concrete bodies sat or lay, for the most part browned and half-naked to the heat. [...] Through and around them others walked tracks that by mysterious consensus remained open; the penned herd seethed with its own impenetrable order.
>
> (169)

It is clear that Alison's language upon viewing the amassed poverty on the streets is racially charged, and therefore requires some unpacking. On the one hand, the language works to reinforce the horror of the overcrowded, disease-stricken conditions of the enclaves. Yet, on the other, Alison's response is obviously predicated on notions of Otherness. Not only are the people dehumanised by their association with livestock, they are also presented as featureless "browned and half naked bodies", and a "packed *desert* of flesh", reinforcing the association between the imagined hostility of the desert and its inhabitants.

Alison's racialised anxieties about the desert are transferred on to the flesh of a homogenised poor, who are subsequently dehumanised. As she moves through the crowd, she thinks that "if one of these ragged dwellers touched me [...] I would scream out in panic" (170). Moreover, here and throughout the novel the Swill are described as dark, dirty, unwashed and related to the floodwater that clings to their clothes and seeps through their apartments. This performs as another racial invocation of indigenous presence. As Toni Morrison (1992) has argued, in the white imagination, African Americans and indigenous peoples are conceptually stained with dirt (vii–xxiii). Dirt, in the form of polluted floodwater, is therefore not considered a concern of the state, but maintains the marginality of the Swill, in part, by what Mike Davis terms "the dialectic of ordinary disaster", whereby disaster is incorporated into history and rendered forgettable specifically because the burden of risk falls unequally on the disregarded poor.

Disasters, such as the incursion of annual flooding, are framed by a ready dismissal from the state and obscured by the media. Indeed, we are told of "Swill communities starving through computer error when Deliveries provided tonnes of salt instead of reduced protein, or some such stupidity, by the newscasts reported only the odd comic mix-up that had no serious consequences" (83). Similarly, when winter flooding hits the Swill tenements and engulfs the Fringe "carrying branches and bottles and dead animals and clutches of nameless flotsam. Probably sewerage overflow" through the abandoned streets of South Melbourne, the news report only highlights "rain in the hills" (66). As Nixon explains in another context: "[w]hat we witness is a kind of fatal bigotry that operates through the spatialising of time, by off-loading risk onto the 'backward' communities that are barely visible in the corporate media" (66).

Indeed, to the present-day reader, the scene is comparable to the destruction that gave witness to the institutionalised classism and racism during Hurricane Katrina. Yet, it is also not much of a stretch of the imagination to read into the scenes of displacement and forced occupation of the Swill enclaves a subliminal reminder of Australia's historic creation of Aboriginal reserves, where many Aboriginal people were forcibly taken in the late nineteenth century. The reserves were withheld from sale or lease by the Crown, and occupied by Indigenous people without any rights to property, or security of tenure. In most cases, the reserves were managed by government officials or church authorities. The enforced movement of Aboriginal and Torres Strait Islander people onto reserves enabled an administratively convenient concentration of populations, allowing the Australian government to intensify legal controls over First Peoples under the false doctrine of *terra nullius* until the 1992 High Court ruling in *Mabo v Queensland (No 2)*. Moreover, Melissa Nursey-Bray Timothy F. Smith and Phil Rist (2019) and others have observed in their study on Australian Indigenous peoples and climate change that the exposure of Indigenous Australians to environmental changes occurring across Australia is "compounded by existing socio-economic disadvantages such as inadequate health and educational services, insufficient infrastructure and limited employment opportunities linked to colonial and post-colonial periods" (473). In this sense, the extended allusion to Swill dispossession and sociopolitical deprivation speaks to the violent oppressions experienced by indigenous voices excluded from the text.

While the trace or haunting of indigenous presence and absence offers a means of conceptualising the veiled histories of colonial traumas, it is also significant because it features as a double of Australia's past and present ambivalent relationship to migration as an extension of its histories of colonialism and indigenous dispossession. In the novel, a third of Australia is set aside for Asian population relocation in 2033. The process leads to the collapse of the already deteriorating national infrastructures as dispossessed and uncompensated Australian landowners are pushed into overcrowded urban areas, adding to the drastic rise in unemployment levels and pressure on state subsistence.

Meanwhile, a desperate effort to make two-thirds of Australia's uninhabitable land habitable again through geoengineering projects causes a chain of unintended environmental consequences that ultimately brings the economy to bankruptcy. While the events of the "Asian Relocation" are told dispassionately by the narrator, its racial terminology reproduces the nationalistic logic of White Australia, while also connoting an older anxiety over Australia's fragility as an "island nation" at threat of invasion from other non-European countries:

> The Swill who believed that things could only get far worse before they failed once and for all to get better. Few had ever seen the countryside, let alone the outback. Desert, drought and flies, wasn't it? So, let the Veets, the Chinks and Indons have it. No place for the white man. Not much of a place for the yellow man, either; two-thirds of Australia has been uninhabitable through millennia, and to that area we admitted him.
>
> (33)

The vision of the outback as "Australia's empty hectares" that have been uninhabitable for millennia repeats the notion of indigenous absence writing Aboriginal people out of the historical record. Moreover, the outright indifference shown by the Swill illustrates the closed system of this post-apocalyptic settler Australia, where even the marginalised and maligned communities of the Swill enclaves reinforce the selective inclusion and exclusion in the remaining space of the settlement.

The attitude of the Swill speaks to the increasingly restrictive attitudes of the Australian public towards Vietnamese refugees, Australia's first officially labelled "boat people", of the mid- to late 1970s. The historian Nancy Viviani (1984) has observed the exaggerated anxieties of politicians and bureaucrats over the first boat arrivals of refugees from Vietnam to show that, contrary to the hyperbolic inflation of numbers in political rhetoric during the period, only 2,059 people arrived by boat from Vietnam, out of a total of 57,770 Vietnamese refugees settled in Australia between the years of 1975 and 1982. As such, the "Asian Relocation" process in *The Sea and Summer* can be viewed as a satirical comment on the increasingly xenophobic tensions in the public and political arena happening during the period of the novel's composition.

Nevertheless, from a present-day perspective it is impossible to read this section and not be reminded of the human rights disaster that recently unfolded on Manus Island in Papua New Guinea (PNG), as a more recent example of the negligence of the Australian government towards refugees and asylum seekers. This is the last we hear about the climate refugees, who, having gone to desperate and risky lengths to survive, become an obstruction to the nation's future, and therefore, we assume, are left to fend for themselves in the desert. The scene connects to Australia's broader history of violence as a settler colonial nation. Such is the longstanding racial disparity found in Australia's migration histories that the historian Ann Curthoys' (2006) has called for the country's histories of migration to be framed as histories of colonisation in acknowledging how entangled these histories are (277).

While these evils long postdate the publication of the novel, the imaginative and physical expulsion of the climate refugees continues to resonate with the invisibility of refugees in Australia's public sphere, which as Ruth Bailint and Zora Simic (2018) suggest is "an invisibility partly imposed by the secrecy surrounding their treatment by official agencies, and partly because they are often living on the margins of Australian society or out of sight in immigration detention centres [and] offshore processing centres" (404). In this way, *The Sea and Summer* forms a dialogue with the ongoing traumas related to the resurgence of constrictive nationalisms and closure of borders in face of migratory and refugee flows. As Lester R. Brown (2011) explains, "One of the defining characteristics of our time is the swelling flow of environmental refugees: people displaced by rising seas, more-destructive storms, expanding deserts, water shortages, and dangerously high levels of toxic pollutants in the local environment" (n.p.). In relation to Turner's homeland, the 2014 IPCC Climate Change Report on "Impacts, Adaptation, and Vulnerability" paints an eerily familiar picture, estimating that "climate change has the potential to change migration flows within Australasia, particularly because of coastal changes (e.g., from the Torres Straits islands to mainland Australia)" (1410). The report predicts that "the impacts of climate change in the Pacific may contribute to an increase in the number of people seeking to move to nearby countries [...] and affect political stability and geopolitical rivalry within the Asia-Pacific region" (1410). In this way, by tracing the connection between the public and political response to refugees of the late twentieth century and now, I hope to have shown the deeper historical currents of colonial-settler violence replete in the Australian environmental apocalyptic imaginary.

Environmental apocalypse in *The Sea and Summer* is represented through the recurrent and evolving phenomenon of sea level rise, which comes to frame the broader conditions of environmental and socio-economic collapse. The flood also helps the reader to navigate the novel's rather complex temporal shifts between the two interchanging narratives. Throughout the novel, rising floodwaters offer an intimation of the temporal setting, yet also gesture towards the deep time of environmental collapse. Andra's difficulty at the end of the novel in representing the complex effects of environmental collapse can be seen to mirror contemporary debates in environmental humanities scholarship on the challenges of imagining environmental crisis in narrative terms. As Andra suggests, this is partially to do with the scale of environmental crisis, which "[i]n the enormous stretch of history to come the Greenhouse years will rate as little more than an unseasonably hot day..." (361). Andra's comment speaks to the difficulty, or indeed impossibility, in uniting the enormous spatial and temporal scales of climate and environmental change with that more suitable to human comprehension. The ecocritic Timothy Clark (2013) argues that the Anthropocene "enacts the demand to think human life at much broader scales of space and time" than previously offered in conventional narrative form (13). Clark suggests ecocritical analysis should interpret narratives on the global scale as well as at the personal and local. Following Clark's suggestion, the thousand-year time frame in *The Sea and Summer* places the human time-spans of the Conway family within a much broader narrative history that incorporates the vastness of planetary and climatic change.

However, as Andra's incomprehension of the environmental crisis affirms, such colossal time-scales may also create a feeling of dislocation in the reader. This is echoed by Andra upon first seeing the ruins of the Swill tenements, where,

> ...he had been deeply affected by the towers and the sense of an immense past thirty of forty meters below the keel, brought face to face in his creative imagination the vastness of changes that had metamorphosed a planet as mindlessly as cosmic eruptions destroyed and created stars.
> (14)

Andra's comments, and the framework narrative of the Autumn People more generally, demonstrate the difficulties of imagining planetary deep-time on a human scale. While the "sunken city" presents the reader with a not-so-subtle admonition against the destructive practices of neoliberal capitalism, the passage also evokes environmental collapse as an evolving phenomenon that exists beyond the purview of the human species in the vastness of universal deep-time.

The passage further subverts the "end of the world" narrative as Lena goes on to admonish Andra's view of the impending ice age as a "second Dark Age", responding,

> Don't be dramatic about history. We're very well equipped to endure a million years of cold. Our ancestors weathered an Ice Age in caves and animal skins, hunting with flint-tipped spears. [...] An Ice Age is no great tragedy – it is in fact the normal state of the planet.
> (14)

Lena presents the coming ice age as not the end of the world, but as a condition of the earth's natural cycle, echoing Andra's thoughts on the vastness of the planet's environmental and climactic metamorphosis over millennia. In this way, *The Sea and Summer* presents a dichotomous narrative that, on one hand, utilises the apocalypse genre to warn against the current course of action in the reader's world in pointing towards a clear linear causality. On the other hand, the novel also proposes a nonlinear temporality in which the earth exists in ongoing and self-organising transformation. As such, Turner puts forward a planetary framework that adheres to a conceptualisation of space and time that extends beyond the strictures of the Anthropocene.

Wai Chee Dimock (2003) describes "planetary time" as "temporal length joined with spatial width" (492). It is "both cumulative and nonsequential" and bears markers of every moment in history concurrently (182). Moreover, as Kelly Frame (2017) suggests,

> [t]he macrocosmic scope of a planetary framework reasserts the importance of ecological history (and futurity). [...] [T]his 'big picture' approach to history observes the fundamental interconnectedness between people, societies and cultures, without eliding the predatory and exploitative nature of some of these interactions.
>
> (4)

From this perspective, the Autumn People's concept of the planetary would see them struggle to understand the exploitative logics of capitalism, with its insistence on ever-expanding production, consumption and profit, that led to environmental collapse in the twenty-first century. At the same time, Turner's conceptualisation of deep-time and the planetary establishes a clear connection between the Autumn People and a quasi-Indigenous notion of time. This is further explicated when we learn that Andra has an unmistakable "strong Aboriginal strain", and that "his dark skin made it necessary [...] to use a heavy Caucasian makeup for most roles" (5). This fleeting and only mention of Indigenous presence in the novel underscores a racialised inscription of Indigenous peoples, which we also see in reference to the "dark skin" of the Swill characters. It is interesting that the novel's only explicit description of Australian Aboriginal Indigeneity is repressed with white Caucasian makeup even in the future idealised state of the Autumn People, suggesting that colonial oppression continues to mask Indigenous presence. This jars with the fact that Aboriginality is connoted positively within the context of the Autumn People as a sustainable and resilient survivor society, despite the fact the novel hints at a continued racism.

I have attempted to show how instances of dispossession, forced relocation and the imaginative work of community expulsion feature in *The Sea and Summer* as reminders of continued of settler-colonial rule despite the absence of explicit Indigenous presence (7). Indeed, *The Sea and Summer* remains conspicuously white, suggesting to the casual reader the same sense of historical amnesia which has furthered colonial-settler privilege in the environmental apocalyptic imaginary. Whyte speaks directly to the non-indigenous writer of apocalyptic fictions when he asserts that "[d]etachment erases the fact that Indigenous peoples everywhere have been through repeated apocalypses" (238). While it is possible to ascertain the trace of imperialism and settler colonialism in the violence endured by the Swill, *The Sea and Summer* ultimately conforms to the problematic notion of repeated apocalypse. I hope in addressing the instances of concealed Indigenous dispossession and *denied* histories in the depiction of the social caste system in *The Sea and Summer*, I have posed a challenge to twenty-first-century environmental apocalypse narratives to be attentive to the veiled histories of settler colonial capital replete in the genre.

References

Anson, April (2017) American Apocalypse: The Whitewashing Genre of Settler Colonialism. *Academia.edu.*

Bailint, Ruth & Zora Simic. (2018) Histories of Migrants and Refugees in Australia. *Australian Historical Studies* 49:3, 378–409.

Baldwin, Andrew. (2017) Postcolonial Futures: Climate, Race and the Yet-to-Come. *ISLE: International Studies in Literature and Environment* 24:2, 292–305.

Bankoff, Greg. (2019) Remaking the World in Our Own Image: Vulnerability, Resilience and Adaptation as Historical Discourses. *Disasters* 43:2, 221–239.

Bracke, Astrid. (2018) "*Collapse.*" *Climate Crisis and the 21st-Century British Novel.* London: Bloomsbury Academic.

Brown, Lester R. (2011) *World of the Edge: How to Prevent Environmental and Economic Collapse.* New York: W.W. Norton & Company, online.

Buell, Fredrick. (2003) *From Apocalypse to Way of Life: Environmental Crisis in the American Century.* New York: Routledge.

Clark, Timothy. (2013) *Ecocriticism on the Edge: The Anthropocene as a Threshold Concept.* London: Bloomsbury.

Colebrook, Claire. (2017) Anti-catastrophic Time. *New Formations* 92, 102–119.

Corbett, Claire. (2017) Nowhere to Run: Repetition Compulsion and Heterotopia in the Australian Postapocalyptic – from 'Crabs' to MadMax beyond the Thunderdome. *Science Fiction Film and Television* 10:3, 329–351.

Curthoys, Ann. (2006) An Uneasy Conversation: Multicultural and Indigenous Discourses. *The Future of Australian Multiculturalism: Reflections on the Twentieth Anniversary of Jean Martin's the Migrant Presence.* Eds. Ghassan Hage and Rowanne Couch. Sydney: Research Institute for Humanities and Social Sciences, 277–294.

Dimock, Wai Chee. (2003) Planetary Time and Global Translation: 'Context' in Literary Studies. *Common Knowledge* 9:3, 488–507.

Frame, Kelly. (2017) 'The Last Place': The Uncanny Australia of David Mitchell. *Antipodes* 31:1, 4–15.

Freud, Sigmund. (2015) *Beyond the Pleasure Principle.* Ed. Ulrike May and Michael Schroter, *Psychoanalysis and History* 17:2.

Garrard, Greg. (2004) *Ecocriticism.* Oxford: Routledge.

Gergan, Mabel, Sara Smith & Pavithra Vasudevan. (2018) Earth beyond Repair: Race and Apocalypse in Collective Imagination. *Environment and Planning D: Society and Space* 38:1, 91–110.

Haynes, Roslynn D. (1998) *Seeking the Centre: The Australian Desert in Literature, Art and Film*. Cambridge: Cambridge University Press.

Intergovernmental Panel on Climate Change. *Climate Change 2014 – Impacts, Adaptation and Vulnerability: Part A: Global and Sectoral Aspects: Working Group II Contribution to the IPCC Fifth Assessment Report*. Cambridge: Cambridge University Press.

Merchant, Carolyn. (2003) Shades of Darkness: Race and Environmental History. *Environmental History* 8:3, 380–394.

Morrison, Toni. (1992) Introduction. *Race-ing Justice, Engendering Power: Essays on Anita Hill, Clarence Thomas and the Construction of Social Reality*. Ed. Morrison, Toni. New York: Random House.

Nursey-Bray, Melissa, R. Palmer, T.F. Smith & P. Rist. (2019) Old Ways for New Days: Australian Indigenous Peoples and Climate Change. *Local Environment* 24:5, 473–486.

Nixon, Rob. (2011) *Slow Violence and the Environmentalism of the Poor*. Cambridge: Harvard University Press.

Seraphin, Bruno. (2016) 'The Hoop' and Settler Apocalypse. *The Trumpeter* 32:2, 126–146.

Turner, George. (2013) *The Sea and Summer*. London: Gollancz.

Weaver, Roslyn. (2006) At the End of the World: Australian Adolescent Literature and Apocalypse. *Journal of Fantastic Arts* 17.2:66, 155–168.

Weaver, Roslyn. (2007) "At the Ends of the World: Apocalypse and Australian Speculative Fiction". *University of Wollongong Thesis Collection*, https://ro.uow.edu.au/cgi/viewcontent.cgi?article=2733&context=theses

Weaver, Roslyn. (2011) *Apocalypse in Australian Fiction and Film: A Critical Study*. Jefferson: MacFarland & Company, Inc.

Whyte, Kyle P. (2018) Indigenous Science (Fiction) for the Anthropocene: Ancestral Dystopias and Fantasies of Climate Crises. *E Nature and Space* 1:1–2, 224–242.

Wynter, Sylvia. (2003) Unsettling the Coloniality of Being/Power/Truth/Freedom: Towards the Human, after Man, Its Overrepresentation—An Argument. *CR: The New Centennial Review* 3.3, 257–337.

Yusoff, Kathryn. (2018) *A Billion Black Anthropocenes or None*. Minneapolis: University of Minnesota Press.

5 Apocalyptic literary geographies

The Tempest's 'brave new world,' *Frankenstein*'s 'modern Prometheus' and *Cloud-Atlas*' 'furthest-seein' eye'

Charles Travis

The philosopher Hannah Arendt (1961 258) noted that the 'political function of the storyteller-historian or novelist is to teach acceptance of things as they are. Out of this acceptance, which can also be called truthfulness, arises the faculty of judgment.' Elements of Arendt's thought will be elicited to explore the apocalyptic literary geographies plotted by William Shakespeare's *The Tempest* (1611), Mary Shelly's *Frankenstein, or The Modern Prometheus* (1818) read in conjunction with Lord Byron's poem *Darkness* (1816) and the intersecting time-travelling narrative arcs of David Mitchell's *Cloud Atlas* (2004). Culturally and historically, the narratives of these works contextualize the Anthropocene in a variety of apocalyptic prisms from its eighteenth- and nineteenth-century origins, to its evolution and twenty-first-century manifestations. The etymology of the word 'apocalypse' can be traced to the Greek word *apokalyptein*, to describe an uncovering of something concealed, that once revealed facilitates a subsequent transformation in perception and consciousness. The word was deployed eschatologically in the *Book of Revelations*, by John of Patmos exiled on the eponymous island in the Aegean Sea during the reign of the Roman emperor Domitian (81–96 CE). In recent popular use, *apocalypse* has often been conflated with the Greek word *Armageddon* meaning the 'end of the world,' and located in biblical prophecy to the ancient city of Megiddo, in today's northern Israel, where a final battle between good and evil is predicted to take place. Similar conflations and imaginaries have also suggested that 'fantasies of the apocalypse are both a product and a producer of the Anthropocene' (Ginn, 2015). In *The Human Condition* (1958), Arendt discusses the phenomena of 'world alienation,' by sketching how three monumental events dragged Western history kicking and screaming from the Dark Ages over the threshold of modernity. First, the discovery of America and the global circumnavigations aimed to 'enlarge the earth,' but which succeeded in 'shrinking her into a ball.' Events on *The Tempest*'s imaginary island surrounded by a storm-tossed sea speak to the turning tides of perception birthed by the age

DOI: 10.4324/9781003128854-6

of discovery from the morass of medieval cosmology to the empirical clarity of the scientific and industrial revolutions.

Second, the processes of resource extraction accompanying the Protestant and Counter Reformation's re-definition of property, the role of the nation state, secularization, capital, commerce and the accumulation of private wealth. Shelley published *Frankenstein, or the Modern Prometheus*, in the aftermath of the French Revolution, and prior to the massive labour appropriations and resource extractions of the industrial revolution. Read in tandem with Lord Byron's poem about ecocide, *Darkness*, Shelley intimates in *Frankenstein* that the re-assemblage of the human condition also possessed a monstrous agency to transform and unleash itself as a geological force – in effect an uncontrollable 'force of nature.'

Third, the invention of the telescope, which opened the 'secrets of the universe,' to 'human cognition,' symbolized a new Galilean ontology of science and technology and refocused 'the earth from the viewpoint of the universe' (Arendt, 1998, 248–261). In such a manner, the six episodes comprising Mitchell's *Cloud Atlas* 'telescope' across space and time rendering a narrative travelogue from the nineteenth-century tropics of the South Pacific Ocean, across twentieth-century Europe and North America, to a futuristic, dystopian Korea and finally to a lush, post-apocalyptic Hawaii where Zach'ry, *Cloud Atlas*' core character, dwells. The novel's principal story relays a tale of the peaceful Valley Folk, the extra-terrestrial Prescients and the atavistic Kona tribe culminating in a visit to the sacred ruins of the Mauna Kea Observatory, a relic of a past and lost 'scientific' civilization.

The literary geographies of 'revelation' in Shakespeare's, Shelley's and Mitchell's texts speak to the overlapping historical, cultural and technological agencies of the human condition which in Arendt's eyes characterized the emergence of global modernity and, thus by extension, the Age of the Anthropocene. While *The Book of Revelation* imparts an enigmatic *apocalyptic* vision, with John the 'Revelator' painting the decline of the pagan Roman Empire and the arrival of the Christian 'kingdom at hand.' In a similar manner, the narratives of the works discussed in this chapter offer radical departures from the spiritual, political and epistemic sensibilities of the cultural milieus in which they were composed.

The publication dates of *The Tempest*, *Frankenstein* (in addition to Byron's *Darkness*) and *Cloud Atlas* can also be seen as cultural benchmarks contextualizing the origins, emergence and evolution of the Anthropocene as a new geological epoch. Collectively, these works provide a speculative vision of how this new age's past and future co-exist and illustrate that 'literature in its hyper-evolved discourse can capture and describe the time-bound, fluctuant, unstable growth of organic life and of human activity in the macroscopia' (Porush, 1991, 372). As Arendt argues in *The Human Condition*, human action is primarily symbolic in character, with the web of human relationships sustained by communicative interaction – not only at an interpersonal level, but through the production of literature, film and art.

By mapping the literary geographies plotted by these writers, distinctions and corollaries can be made between the Apocalyptic and Anthropocenic foci of their works.

The Tempest (1611)

> . . . but Nature should bring forth,
> Of its own kind, all foison, all abundance,
> To feed my innocent people.
>
> *The Tempest* (Act 2, Scene 1, 178–180)

It has been argued that Shakespeare's island set drama *The Tempest* was inspired in the seventeenth century by English conflict with the Ottoman Turks in the eastern Mediterranean, William Strachey's 1609 account of a Bermuda shipwreck, in addition to the chartering and settlement of the Virginia Colony between 1606 and 1607. However, Shakespeare's play can also be contextualized by the establishment in 1610 of an English Grand Banks fishery colony on Newfoundland island, which became the anchor of the North Atlantic Cod and Sack trade triangle (Travis et al., 2020). *The Tempest* premiered a year later on 'Hallowmas nyght,' on 1 November 1611 in the candlelit space of the Blackfriar's Theatre in London. One imagines the oceanic drama opening to the 'tempestuous noise of thunder and lightning heard,' on a deck of ship, on a stage surrounded by 'the cheeky wind blowers of sixteenth and seventeenth century world maps' (Brayton, 2012, 180). The play tells the story of Prospero, the rightful Duke of Milan ousted by his brother Antonio, and exiled on an island along with his daughter Miranda; Prospero is attended to by servants: a spritely spirit named Ariel and Caliban, an inhabitant of the isle who is rendered as 'legged like a man, and his fins like arms!' (*The Tempest*: Act 1, Scene 2, 34–35).

Using artistic and scientific powers of enchantment, Prospero orchestrates a storm, causing the shipwreck of a vessel upon which his brother Antonio, Alonso the King of Naples, his son Ferdinand and court are sailing in return from Tunis after the wedding of the king's daughter Claribel. The survivors emerge from the sea in separate groups and straggle their way to Prospero's abode. Fernando the first to appear from the wreck is imprisoned but set free by the smitten Miranda. A fanciful assassination plot is devised by Caliban, when he is discovered among the flotsam and jetsam of the island's beach by the jester Trinculo and drunken butler Stephano, who escaped the wreck by a 'butt of sack which the sailor's heaved o'erboard' (*The Tempest*: Act 2, Scene 2, 125–126). But Prospero thwarts the plot, with the aid of Ariel, and conjures up a series of magical and alchemical encounters between the shipwrecked that leads to the restoration of his dukedom and the marriage of Miranda and Ferdinand.

The writing and staging of *The Tempest* in 1611 coincided with the establishment of a fishery colony on Newfoundland island in 1610. Fish stalls were abundant in Shakespearean London, and until the 'Sack and Cod' trade shipped cod fished from the Grand Banks and dried on the rocky shores of Newfoundland directly to ports in France, Portugal and Spain in the 1630s, the Thames served as a major trading hub and watercourse to these European fishery markets (Test, 2008). Salted Atlantic Cod (*Gadus Morhua*) is referenced directly as 'Poor John" in *The Tempest* and provided a staple of cheap protein for patrons of the Blackfriar's and adjacent Mermaid's Tavern. The theatre and ale-house were located in a district of London where the 'commodities of plays and fish' were ubiquitously consumed (Brayton, 2012, 159). Indeed, Shakespearean London, overlapped with the emergence of the modern commercial city, capitalized in part by international trade in 'Poor John,' harvested from the Grand Banks.

> MIRANDA: O wonder!
> How many goodly creatures are there here!
> How beauteous mankind is! O, brave new world,
> That has such people in 't!
> PROSPERO: 'Tis new to thee.
>
> *The Tempest* (Act 5, Scene 1, 215–219)

The foundations for the Newfoundland English Oceanic Plantation were laid when the Genoa sea-explorer Giovanni Caboto (John Cabot) sailed into the port of Bristol, England in August 1497 bearing news of a 'New World' cod fishery in the north-west Atlantic 'swarming with fish, which can be taken not only with the net, but in baskets let down with a stone' (Lawrence et al., 1987, 274). In 1583, Sir Humphrey Gilbert declared English possession of Newfoundland at St John's Harbor, before a fleet of over 30 fishing ships from Spain, Portugal, France and England.

In 1610, the London and Bristol-based Newfoundland Company incorporated the Grand Banks as a commercial offshore fishery. Between 1604 and 1620, more than 33% of English MPs invested funds in imperial enterprises such as the trade in Grand Banks cod. The Newfoundland island colony and its waters created a source of capital that allowed Protestant England to conquer the Catholic Spanish Armada in 1588 and counter the ambitions of the Holy Roman Empire in Europe while also supplying cod to its Lenten cultures. As the noble Gonzalo, advisor to the King of Tunis in *The Tempest*, remarks,

> Had I plantation of this isle, my lord -[. . .]
> I' th' commonwealth I would by contraries
> Execute all things, for no kind of traffic
> Would I admit; no name of magistrate;
> [. . .]
> I would with such perfection govern, sir,
> T' excel the Golden Age.
>
> *The Tempest* (Act 2, Scene 1, 157–184)

One post-colonial reading of Caliban's native claim and description of the island in *The Tempest* is perhaps an allusion to English colonial encounters with Newfoundland's indigenous Beothuk people. But as Prospero informs Miranda, Caliban 'serves in offices / That profit us' (*The Tempest*: Act 1, Scene 2, 374–375). His claim on the island could also be read as an allusion to the free and indentured fishery labour pool drawn from the Thames districts and the wests of England and Ireland who created the wealth for fishery investors in London and Bristol:

> CALIBAN
> This island's mine by Sycorax, my mother,
> Which thou tak'st from me. When thou cam'st first
> Thou strok'st me and made much of me; wouldst give me
> Water with berries in't, [. . .]
> And showed thee all the qualities o' th' isle,
> The fresh springs, brine pits, barren place and fertile.
> [. . .] Which first was mine own king; and here you sty me
> In this hard rock, whiles you do keep from me
> The rest o' th' island.
>
> *The Tempest* (Act 1, Scene 2, 396–405; 409–411)

By the eighteenth century, the Grand Banks fishery had evolved into a major early modern geo-strategic theatre of conflict pitting the French against the British. In 1719, the French Jesuit historian Charlevoix contented that the Grand Banks' cod fisheries were the 'true mines, which are the more valuable, and require much less expence [sic] than those of Peru and Mexico' (Roberts, 2010, xxvii). In June 1762, as the Seven Year's war was reaching it conclusion, the French captured St John's Newfoundland, burning English vessels and destroying its fishery infrastructure. However, coinciding with an anonymous restaging of the drama under the title *The Tempest of Enchanted Island* in September, the British counter-attacked (Taylor, 2012). By recapturing Newfoundland from the French forces, a prophetic resonance was given to the lines proclaimed by the restored Duke of Milan at the close of the play's revival:

> PROSPERO
> Weak masters though ye be, I have bedimmed
> The noontide sun, called forth the mutinous winds,
> And 'twixt the green sea and the azured vault
> Set roaring war; to the dread- rattling thunder
> Have I given fire
>
> *The Tempest* (Act 5, Scene 1, 50–55)

The Globe Theatre burned down in 1613 and there is no record of *The Tempest* being performed on its stage. However, the theatre's name and the numerous performances of Shakespeare's dramatic literary geographies on the banks of the Thames embody and emplace Arendt's (1998, 251) observation that: 'before we knew how to circle the earth, how to circumscribe the sphere of human habitation in days and hours,' Elizabethan globe maker and Shakespeare's contemporary Emery Molyneux's use of 'numbers, symbols and models' scaled the physical distances of the earth 'down to the size of the human body's natural sense and understanding.' Contextualizing Shakespeare's dramatic canon with Arendt's discussion of the geographical insights gained during the 'age of discovery' coupled with the ascendancy of the English oceanic nationalism, and its imperial navy, for which the Grand Banks Newfoundland fishery was certainly a catalyst, illustrates that 'there is a larger sense in which *The Tempest* is a prologue to the whole thrust of technological modernity' (Bate, 2013, 77). Shakespeare's drama was published the same year in which John Donne's (1611) poem *An Anatomy of the World*, shuddered at the impact from the 'revelations' of the new modern scientific perspective 'invented' by Galileo and Copernicus, which would be translated into Enlightenment rationalism by Descartes, Newton and others:

> And new Philosophy calls all in doubt,
> The Element of fire is quite put out;
> The Sun is lost and th'earth, and no mans wit
> Can well direct him where to looke for it . . .

Despite Donne's protest, the Enlightenment's eighteenth-century 'Clockwork Universe' within which a Cartesian globe (modelled by Molyneux) soon orbited in 'absolute' sequential time. It was based upon rules established by Copernicus' cosmology, Newton's physics, and a grid of coordinates conjured by Descartes' marriage of Euclidian geometry and algebra. The alchemical confluence of the emergence of Western modernity and Cartesian dualism conspired in the 'radical splitting off of the mind from the human body and the body of the earth' (Tally and Battista, 2016) causing the latter to become an ecological subaltern.

As a result of this global *res cogitans* the early modern consumption of Grand Banks cod, read through the dramatic prism of *The Tempest*, serves a harbinger of the Anthropocene, and its first manifestation in steam, coal and the fossils harvested and burned as fuel for the industrial revolution vomiting rivers of CO_2 into the sky, raising the temperature of the atmosphere, creating cascading climatic and environmental effects that spread unevenly across the face of our planet. This chapter's reading of *The Tempest* suggests how the alchemy of literature, geography and the Anthropocene produces reflections expressed as a new type of 'magical environmentalism.' Shakespeare's apocalyptic drama signals the end of one era, and the dream which will imagine the beginning of another. This is audaciously intimated in *The Tempest*'s somnolent coda:

> . . . the great globe itself,
> Yea, all which it inherit, shall dissolve,
> And, like this insubstantial pageant faded,
> Leave not a rack behind. We are such stuff
> As dreams are made on, and our little life
> Is rounded with a sleep.
>
> *The Tempest* (Act 4, Scene 1, 169–175)

Indeed, intimations of a collapse of the Grand Banks cod fishery were heard as 'as early as 1703 [when] an Englishman lamented that 'the fish grows less, the old store being consumed by our continual fishing' (Bolster, 2006, 568). Just over two centuries later in July of 1992, the Canadian government peering into an 'ichthyological apocalypse' and fearing the extinction of Atlantic Cod placed a moratorium on five centuries of Grand Banks fishing. The dissolution of the fishery resulted in the single largest mass layoff in Canadian history, its effect rippling through north-east Canadian and American communities, fishing crews, workers in processing plants, global fishmongers and shipyards (Organization for Economic Co-operation and Development, 2011).

Frankenstein, or the Modern Prometheus (1818)

Shelley opens *Frankenstein* as an epistle in St Petersburg, Russia that relays the excitement of an impending expedition to the Arctic Circle. The novel which will be discussed as the re-assembly of the human condition closes in tragedy with ice floes cracking like thunder across the frozen wastes of the Arctic Ocean. The expedition story framing Shelley's novel was influenced by the comedic, romantic and mystical nature of John Barrow's 1816 Arctic account published in the journal *Quarterly* in February 1817. But it was during the 'wet, ungenial' weeks of June and July 1816, described as the 'Year without a Summer,' that Shelley 'created' her Frankenstein monster.

Global temperatures had plummeted due to the 1815 eruption of Mount Tambora in the Dutch-East Indies (Indonesia) and the amount of particulate matter ejected into the earth's atmosphere clouded the sun's rays. Visiting Switzerland in 'the environs of Geneva' and Shelley and her husband Percy became the neighbours of the infamous Lord Byron. The volcanic induced mood of Shelly's *Frankenstein* can be teased out with a contrapuntal reading of excerpts from Byron's poem *Darkness* (1816) which captures the gloomy summer atmosphere cloaking the Alpine lakeside as a poetic dirge on the teleology of human-environmental relations. Byron first exhibited a growing environmental consciousness *Childe Harold's Pilgrimage* which relays impressions gained during his 1810–1811 tour of Greece (Hubbell, 2010).

Byron's poetic perception of the palimpsestic nature comprising a location's 'sense of place' expanded in direct proportion to the site's particular geo-historical significance. He 'wrote about places, about specific "spots" as if they were locales in which direct connection with the buried subjectivity of the lived experiences associated with those places was somehow (supernaturally) possible' (Cheeke, 2002, 142). Commencing *Darkness*, Byron (1816) echoes both Shelley's method of apocalyptic inspiration, and the bleak arctic foci with which she began to imagine and eventually came to structure her novel:

> I had a dream, which was not all a dream.
> The bright sun was extinguis'd, and the stars
> Did wander darkling in the eternal space,
> Rayless, and pathless, and the icy earth
> Swung blind and blackening in the moonless air;
> Morn came and went -and came, and brought no day . . . .

During the rainy holiday, Byron proposed to Mary and Percy that they all should write stories about a 'supernatural occurrence' to while away the rainy days (Shelley, 1994, 9). At first a bit flummoxed by the task, Shelley took to bed one night at the 'witching hour' after hearing talk of electrical experiments in 'galvanism' during which a Dr Darwin induced voluntary movement in 'a piece of vermicelli in a glass case' (Shelley, 1994, 9). For Arendt, 'dreaming' provides discursive and ideological contexts within which to muse on the potentials of new technologies to potentially transform the human condition (Kearnes, 2007). Reflecting on whether a human corpse could be reanimated in a similar manner, an image of Victor Frankenstein assembling expropriated body parts in the university town of Ingolstadt, Bavaria, came to Shelly in a manner similar to the opening lines of Byron's *Darkness* – in a dream-like fugue state:

> . . . I saw the pale student of unhallowed arts kneeling beside the thing he had put together. I saw the hideous phantasm of a man stretched out, and then, on the working of some powerful engine, show signs of life . . . to mock the stupendous mechanism of the Creator of the World . . . He sleeps but is awakened; he opens his eyes; behold the horrid thing that stands at his bedside, opening his curtains and looking on him with yellow, watery, but speculative eyes.
>
> (Shelley, 1994, 9)

Frankenstein's galvanic resurrection of a corpse leads to a deadly travelogue from Switzerland, to above the Arctic Circle. The trek leaves a wake of deaths that spreads like a plague across the geography of western and northern Europe: Victor's younger brother William is strangled by the creature on the plains of Plainpalais; the subsequent execution of the innocent servant Justine Moritz, framed for the boy's murder by the creature, occurs in Geneva; Frankenstein's fiancé Elizabeth is murdered at Evian on the night of their wedding; his friend, Clerval, is killed in Ireland. Indeed, Frankenstein's creature, though re-animated human flesh, seems to manifest as a force of nature. In Byron's *Darkness* (1816), this geography of subtle terror slowly diffuses across an apocalyptic landscape:

> And men forgot their passions in the dread
> Of this their desolation; and all hearts
> Were chill'd into a selfish prayer for light:
> And they did live by watchfires -and the thrones,
> The palaces of crowned kings -the huts,
> The habitations of all things which dwell,
> Were burnt for beacons; cities were consum'd,
> And men were gather'd round their blazing homes
> To look once more into each other's face; . . . .

In *Frankenstein*, after Justine's execution, Victor pursues the monster into the Alps and across the glacier 'rising like a troubled sea' between Montanvert and Mont Blanc:

> . . . suddenly I beheld the figure of a man, at some distance, advancing towards me with superhuman speed. He bounded over the crevices in the ice, among which I had walked with caution . . . it was the wretch whom I had created . . . its unearthly ugliness rendered it almost too horrible for human eyes.
>
> (Shelley, 1994, 94–95)

In the cold, aerie of the Alps, the creature begins to tell Frankenstein his own story, from the first day of his consciousness to their encounter on the glacier. Shelley frames the creature as 'a "natural man" who recapitulates the stages between man in the state of nature and man in civilized society' (Ozolins, 1975, 105). The creature readily informs Frankenstein that:

> The desert mountains and dreary glaciers are my refuge . . . the caves of ice which I do not fear are a dwelling for me . . . if the multitude of mankind knew of my existence, they would do as you do and arm themselves for my destruction.
>
> (Shelley, 1994, 96)

In contrast to Prospero's acknowledgement in *The Tempest* that Caliban is this '"thing of darkness I acknowledge mine" – Victor's reaction to the creature is "first to flee, then to kill"' (Ozolins, 1975, 104). Hoping to end the murders, Frankenstein, at the creature's request, retreats to the barren Orkney Islands to stitch him together a mate. Byron (1816) writes that '*Even dogs assail'd their masters*' and his poem echoes the hollowness of Victor's false optimism:

> Happy were those who dwelt within the eye
> Of the volcanos, and their mountain-torch:
> A fearful hope was all the world contain'd;...

However, Shelley (1994, 165) writes that the 'remains of the half-finished' subject mangled from 'the living flesh of a human being' brings Victor only revulsion, and he tosses the body parts into the Atlantic and resolves to hunt his creation down. The last meeting between the creature and Frankenstein is at the latter's deathbed in the freezing cabin of a vessel stranded in the Arctic ice floes. The frigid image is juxtaposed by Shelley with the monster's imminent self-immolation:

> I will ascend my funeral pile triumphantly and exult in the agony of the torturing flames. The light of that conflagration will fade away; my ashes will be swept into the sea by the winds. My spirit will sleep in peace.
>
> (Shelley, 1994, 215)

The text of *Frankenstein* lets the reader ponder ideas about self and nature, ethics and how Western modernity and Cartesian dualism conspire to create an abyss between the mind of the human body and the body of the earth. Intimations of this despair which haunt the creature and its creator are captured in Byron's (1816) poem:

> Their funeral piles with fuel, and look'd up
> With mad disquietude on the dull sky,
> The pall of a past world; and then again
> With curses cast them down upon the dust,
> And gnash'd their teeth and howl'd: the wild birds shriek'd
> And, terrified, did flutter on the ground,
> And flap their useless wings; the wildest brutes
> Came tame and tremulous; . . . .

Endeavouring to 'preserve the truth of the elementary principles of human nature' in her terror tale, Shelley (1994) listed her influences as the *Iliad*, the tragic poetry of Greece, Shakespeare's *The Tempest* and especially Milton's *Paradise Lost.* Shelley retired to the old Roman colony of Bath, in England to write her version of *Prometheus Unbound.* In distinction to Shakespeare's rending of Prospero's alchemical powers, the magical Ariel and the 'beastly' Caliban, 'Frankenstein's "being" is *the* monster of modernity. As the first novel in the genre science fiction stories, Frankenstein marked a departure from traditional ghost stories or tales of the supernatural to stories about the potential horrors of science and technological development' (Hammond, 2004, 184). James Lovelock (2019, 33) notes as the modern age dawned, the increase in population, multiplication of nation states, bourgeoning empires and subsequent wars led to a colossal demand for wood: 'at the beginning of the eighteenth century the building of a warship could consume 4,000 trees – and for smelting iron. Forests were being depleted faster that they could regrow.' In *Darkness* (1816), Byron depicts this massive ecocide:

> Forests were set on fire -but hour by hour
> They fell and faded -and the crackling trunks
> Extinguis'd with a crash -and all was black . . . .

However, with the dawn of the Anthropocene, the extraction of fossil fuels would supersede the exploitation of forests. Like Prometheus stealing fire from the gods, the geologically stored energy of the sun would now come to be consumed at an obscenely higher exponential rate and scale. For Lovelock (2019, 33), the steam pump to which Thomas Newcomen (1663–1729) was attributed in *Monthly Chronicle* as 'sole inventor of that surprising machine for raising water by fire' was the first significant piece of Anthropocene technology. 'This little engine' according to Lovelock (2019, 34) 'did nothing less than unleash the Industrial Revolution,' by pumping water from flooded mines and hastening the extraction of coal. As Will Steffen et al. (2011, 848) observe, the new technologies of the industrial revolution caused human energy consumption to rise sharply with the new society of the Anthropocene using 'four or five times as much energy as their agrarian predecessors, who in turn used three or four times as much' as their 'hunting and gathering forebears.' Like an Old Testament prophet, Byron laments in *Darkness* (1816) the birth of this new extraction-based society:

> For an unholy usage; they rak'd up,
> And shivering scrap'd with their cold skeleton hands
> The feeble ashes, and their feeble breath
> Blew for a little life, and made a flame
> Which was a mockery; . . .

Returning to *Frankenstein*'s 'initial plotting' it can be seen that Shelley 'systematically places her Gothic horrors within the geographical and political particularities of European and world history,' depicting 'forces that cannot be confined by the political control or geographic space' (Randel, 2003, 487–488). The climatic landscapes of the 1789 French Revolution (precipitated by drought-induced grain riots) and Napoleon's disastrous 1812 Winter Campaign to Moscow (marked by precipitous drops in temperature) inform *Frankenstein*'s tropes. The role of climate in European sociopolitical conflict is intimated in Byron's (1816) verses:

> And War, which for a moment was no more,
> Did glut himself again: a meal was bought
> With blood, and each sate sullenly apart
> Gorging himself in gloom: no love was left;

All earth was but one thought -and that was death
Immediate and inglorious; and the pang
Of famine fed upon all entrails -men
Died, and their bones were tombless as their flesh; . . . .

Frankenstein anticipates how the emergence of the modern age and the mechanistic nature of the industrial revolution, and its ravenous consumption of fossil fuels would come to symbolize the monstrous potential of a new and constructed human agency to unleash itself as a natural, geological force by inaugurating the age of the Anthropocene. In response, the concluding verses of Byron's *Darkness* (1816) speak to the primordial agency and emergence of a conscious universe.

The winds were wither'd in the stagnant air,
And the clouds perish'd; Darkness had no need
Of aid from them – She was the Universe.

Anticipated by John Barrow and Frank Tippler's *The Anthropic Cosmological Principal* (1986), the cosmic consciousness of the universe alluded to in Byron's poem will be birthed according to Lovelock (2019) by the confluent processes of the Anthropocene and Gaia acting as human and planetary midwives. *Darkness*, though rooted in the gloomy vista of Lake Geneva during the summer of 1815, is similar in narrative latitude to *Frankenstein*, as it takes on a pan-European, global and almost cosmic scale in its perceptive scope of the cultural and apocalyptic dimensions and re-assembled human condition spawned by the genesis of the Anthropocene.

Cloud Atlas (2004)

It is on an island in the tropical South Pacific Ocean in the 1850s that the first of *Cloud Atlas*' five episodes inaugurates the novel's time and space travel to Belgium in the 1930s; the west coast of America in the 1970s; Britain in the early 2000s; a speculative future of the *Nea So Copros* on the Korean Peninsula and finally the post-apocalyptic Big Island of Hawaii. It is from this location that novel's core story *Shoosha's Crossin' an' Ev'rythin After*, narrated by the 'Valley Man' Zach'ry, occurs before *Cloud Atlas* circles back in time and space through the codas of the five preceding episodes to close in the South Pacific nineteenth century.

Anglo-American readings of the Mitchell's six inter-nested episodes are somewhat complicated by variances in the storylines of chapters in the British and US editions of *Cloud Atlas*.[1] Adapted by film directors Tom Tykwer and the Wachowski brothers (him/hers/they), the cinematic version of the novel finds Zach'ry 'narrating' the *Cloud Atlas* in *patois* on a distant planet as he provides an oral history of the post-apocalyptic earth. 'Yarnin' to his 'babbits' Zach'ry tells the story of Meronym who arrived on a 'Great Ship' of the 'Prescients' to study the Valley Folk and their enemies, the Kona, a slave-trading, warrior tribe. Meronym asks Zach'ry to guide her up to the sacred and forbidden summit of the Big Island's volcano where the ruins of the Mauna Kea Observatories ('where the Old Uns worshiped their Smart') are perched, and 'Old Georgie,' the devil in the Valley Folk's mythology, lurks (Mitchell, 2004, 275).

The Observatory's transition from an Arendtian symbol of 'science' to a 'post-apocalyptic' sacred space can be parsed through the lens of eighteenth-century Italian political philosopher Giambattista Vico's cyclical perspective on history. In *Scienza Nuova* (1725), Vico posited that the development of societies ascended through three distinct ages: *Theocratic*, *Aristocratic* and *Democratic*, before '*la barbarie della riflessione*' (the barbarism of reflection), unleashes an age of demagoguery, social anomie, anarchy and chaos, prompting a *Ricorso*, the collapse of civilization, and inaugurates a new cycle of history, ruled by oracles and gods (Vico, 2013). Zach'ry's description of the '*obser'trees*,' as 'temples' with 'white'n'silvry an' gold'n'bronz' where the 'Old Uns used to study the planets'n'moon'n'stars, an' the space b'tween, to und'stand where ev'rythin' begins an' where ev'rythin ends,' renders a Viconian ontological shift to Manua Kea's cultural role:

> Shimm'rin floors, white walls'n'roofs, one greatchamber, round'n'sunk, filled by a mighty tube wider'n a man an' longer'n five what Meronym named a *radyo tel'scope* what was, she said, the furthest-seein' eye the Old Uns ever made.
>
> (Mitchell, 2004, 275–276)

Zach'ry's post-apocalyptic oral history is the centrifugal hub around which the spokes of *Cloud Atlas*' other five narratives revolve: *The Pacific Journal of Adam Ewing* (1850s: the colonial exploitation of South Pacific islanders), *Letters from Zedelghem* (1930s: composition of the *Cloud Atlas* symphony in Belgium), *Half-Lives: The First Luisa Rey Mystery* (1970s: the uncovering of the malfeasance of a California nuclear power company), *The Ghastly Ordeal of Timothy Cavendish* (early 2000s: the escape of an involuntary committed London book editor from the confines of an old age 'care home') and *An Orison of Sonmi~451* (a dystopian Korean future, where a cyborg-fabricant named Omni recounts the consciousness of 'being' before her termination).

Excepting Zach'ry, each episode's characters share a comet-shaped birthmark located between their shoulder blade and collarbones. Each episode contains its own unique historical voice, relayed in period English vernacular (or *patois* in the case of Zach'ry) – all coalescing in a polyvocal but legible *Tower of Babel* novel. However, the lacuna in *Cloud Atlas*, hanging over the five ascending and descending staircase narratives to and from the *aerie* of Zach'ry's core post-apocalyptic tale, is the teleology of how his world on the Big Island of Hawaii came to be. Unlike Lovelock's (2019, 104) rosy proclamation in *Novacene* that Gaia will evolve computer based hyper-intelligence systems to watch over our re-assembled human condition with, as the poet Richard Brautigan phrased 'machines of loving grace,' the *An Orison of Sonmi~451* episode hints at the darker outcome relayed by Zach'ry's yarn to his 'babbits.' Fascinated by the '*Smart*' of the '*Old Uns*,' who '*got ev'rythin*' and used to inhabit and dominate the planet Zach'ry is told by Meronym – '*Yay*' but they didn't master '*a hunger for more*':

> . . . more gear, more food, faster speeds, longer lifes, easier lifes, more power, yay. Now the Hole World is big, but it weren't big 'nuff for that hunger what made Old Uns rip out the skies an' boil up the seas an' poison soil with crazed atoms an' donkey 'bout with rotted seeds so new plagues was borned an' babbits was freak-birthed. Fin'ly, bit'ly then quicksharp, states busted into bar'bric tribes an' the Civ'lize Days ended, 'cept for a few folds'n'pockets here'n'there, where its last embers glimmer.
>
> (Mitchell, 2004, 272–273)

In late twentieth-century 'Cold War' imaginaries, the roads to Armageddon commenced with the detonation of the respective atomic bombs 'Little Boy' and 'Fat Man' over Hiroshima and Nagasaki in 1945, showcasing the full Promethean horror of the Anthropocene to the planet's inhabitants. The twin 'unforgettable fires' at Hiroshima and Nagasaki literally and figuratively embedded indelible imprints of anthropic elements created by human-induced nuclear fission within the sedimentary layers of the earth and within the cultural memory of Gai's consciousness perhaps seeding a planetary mental illness. And there our nuclear spoor rests and metastasizes, waiting for the 'Prescients' and their light-year Geiger-counters, like the fossilized remains of a paleolithic fish skeleton, or the gentle tendrils of a pre-historic flora anticipated their discoveries by Victorian antiquarians.

Conclusion

The magical alchemist Prospero, Victor Frankenstein's nameless and re-assembled creature, and the Valley Folk tribesman Zach'ry occupy imaginary landscapes bridging the Anthropocene with its abysmal and liberating apocalyptic potentials. The articulations of these three characters, scripted respectively by Shakespeare, Shelley and Mitchell, flesh out the historical and cultural agencies and literary geographies that illustrate Arendt's three tropes on the birth of Western modernity: (1) discovery and ability to 'shrink the world' – *The Tempest*; (2) expropriation and resource extraction – *Frankenstein* (parsed by Byron's *Darkness*) and (3) the fallacy of the 'far-seeing eye' of science – *Cloud Atlas*. Since the dawn of modernity, the planet and non-*homo sapiens* species have been relegated to the status of ecological subalterns and treated as objects and sites of exploitation, massive appropriation and resource extraction.

The birth of the Anthropocene heralded Armageddon for Atlantic Cod, and the Passenger Pigeon, both sources of cheap protein to feed the labouring masses and for the Beaver and the American Bison, creatures eviscerated from their pelts and hides to clothe the owners of the means of Industrial Age production. Early-modern greed for Meso-American gold and silver by bankrupt 'Old World' empires was mimicked by our Modern Age's appetite for coal, ore, iron, fossil fuel and uranium deposits to fuel American, European and Asian Industrial, Automotive, Aeronautical and Atomic political-economic and technological revolutions. This chapter has addressed questions that can be framed in part by paraphrasing *The Second Coming*, a 1919 poem by William Butler Yeats. Do the apocalyptic literary geographies of *The Tempest*, *Frankenstein*, *Darkness* and *Cloud Atlas* voice humanistic, idiosyncratic, radical visions that trace the birth of the 'rough beast' of the Anthropocene – 'its hour come round at last,' as it 'Slouches towards Bethlehem to be born?' (Yeats, 1919). The answer is emphatically yes.

The apocalyptic visions of these pieces of literature when parsed by the Greek root *apokalyptein* reveal how Enlightenment valorization of rationalism, divorced from the vagaries of our re-assembled human condition, characterized the birth of modernity and transformed scientific knowledge from being perceived as a Utopian instrument to harness, a control nature, to a para-dystopian unwieldy and predictable tool whose deployment has facilitated unprecedented consumption of resources, and unforeseen planetary backlashes in the guises of global warming, extreme weather events and the ravages of the COVID-19 pandemic. In this conclusion, Arendt's acerbic late twentieth-century comment on soothsayers, prophets and positivists seems prescient as we ponder our relation with the planet by parsing the meanings and interactions between our understandings of the Apocalypse and the Anthropocene: 'Men, forever tempted to lift the veil of the future-with the aid of computers or horoscopes or the intestines of sacrificial animals-have a worse record to show in these "sciences" than in almost any other scientific endeavor' (1971, 159).

Note

1 The author's first reading of David Mitchell's text was the British Edition of *Cloud Atlas*, but this chapter draws upon the text of the American edition.

References

Arendt, H. (1971) *The life of the mind.* London: Harcourt.

Arendt, H. (1998) *The human condition.* Chicago and London: The University of Chicago Press.

Arendt, H. (1961) *Between past and future: Eight exercises in political thought.* New York: Penguin.

Barrow, J.D., and Tipler, F.J. (1986) *The anthropic cosmological principle.* Oxford: Clarendon.

Bate, J. (2013) *Romantic ecology (Routledge revivals): Wordsworth and the environmental tradition.* London: Routledge.

Bolster, J.W. (2006) Opportunities in marine environmental history. *Environmental History, 11*(3), pp.567–597.

Brayton, D. (2012) *Shakespeare's ocean: An ecocritical exploration*. Virginia, VA: University of Virginia Press.

Byron, Lord (George Gordon) (1816). Darkness. *Poetry Foundation* [online] Available from: https://www.poetryfoundation.org/poems/43825/darkness-56d222aeeee1b

Cheeke, S. (2002) Geo-history: Byron's beginnings. *European Journal of English Studies*, *6*(2), pp.131–142.

Donne, J. (1611) An anatomy of the world. *Poetry Foundation* [online] Available from: https://www.poetryfoundation.org/poems/44092/an-anatomy-of-the-world

Ginn, F. (2015) When horses won't eat: Apocalypse and the Anthropocene. *Annals of the Association of American Geographers*, *105*(2), pp.351–359.

Hammond, K. (2004) Monsters of modernity: Frankenstein and modern environmentalism. *Cultural Geographies*, *11*(2), pp.181–198.

Hubbell, J.A. (2010) Byron's cultural ecology. *European Romantic Review*, *21*(2), pp.183–203.

Kearnes, M. (2007) (Re)making matter: Design and selection. *Area*, *39*(2), pp.143–155.

Lawrence, A.W., Young, J., and Young, J.I. eds., (1987) *Narratives of the discovery of America*. Maryland, MA: Heritage Books.

Lovelock, J. (2019) *Novacene: The coming age of hyperintelligence*. London: Penguin, Random House.

Mitchell, D. (2004) *Cloud atlas*. New York: Random House.

Organization for Economic Co-operation and Development. 2011. *Fisheries: The lessons of the Grand Banks, observer no 284*. Retrieved from http://oecdobserver.org/news/fullstor

Ozolins, A. (1975) Dreams and doctrines: Dual strands in "Frankenstein". *Science Fiction Studies*, *2*(2), pp.103–112.

Porush, D. (1991) Prigogine, chaos, and contemporary science fiction. *Science Fiction Studies*, *18*(3), pp.367–386.

Randel, F.V. (2003) The political geography of horror in Mary Shelley's. "Frankenstein". *Elh*, *70*(2), pp.465–491.

Roberts, C. (2010) *The unnatural history of the sea*. Washington, DC: Island Press.

Shelley, M. (1994 [1818]) *Frankenstein, or the Modern Prometheus*. London: Penguin Classics.

Steffen, W., Grinevald, J., Crutzen, P., and McNeill, J. (2011) The Anthropocene: Conceptual and historical perspectives. *Philosophical Transactions of the Royal Society A: Mathematical, Physical and Engineering Sciences*, *369*(1938), pp.842–867.

Tally, R., and Battista, C. (2016) *Ecocriticism and geocriticism: Overlapping territories in environmental and spatial literary*. London: Palgrave Macmillan.

Taylor, D.F. (2012) The disenchanted island: A political history of "The Tempest", 1760–1830. *Shakespeare Quarterly*, *63*(4), pp.487–517.

Test, E.M. (2008) The tempest and the newfoundland cod fishery. In Eds. Sebek, B., and Deng, S. *Global Traffic*. New York: Palgrave Macmillan, pp.201–220.

Travis, C., Ludlow, F., Matthews, A., Lougheed, K., Rankin, K., Allaire, B., Legg, R., Hayes, P., Breen, R., Nicholls, J., Towns, L., and Holm, P. (2020) Inventing the Grand Banks: A deep chart. *Geo: Geography and Environment*, *7*(1), pp.1–22. https://rgs-ibg.onlinelibrary.wiley.com/doi/full/10.1002/geo2.85

Vico, G. (2013) *New world encyclopedia* [online]. Available from: http://www.newworldencyclopedia. org/entry/Giambattista_Vico

Yeats, W.B. (1919) The second coming. *Poets.org*. [online] Available from: https://poets.org/poem/second-coming

6 A world without bodies

Geotrauma and the work of mourning in Jorie Graham's *Fast*

Philip Jones

Mourning in the age of climate change

At the beginning of March 2019, the United Nations released a report on levels of warming in the Arctic. According to Schoolmeester et al. (2019), if warming is kept to a medium-emission scenario, temperatures over the Arctic Ocean could be expected to rise between three and five degrees Celsius by mid-century with further increases of five to nine degrees Celsius after this relative to 1986–2005 levels. As reported in *The Guardian* newspaper (Harvey 2019), this would mean reaching a climate 'tipping point', where melting permafrost releases previously trapped greenhouses gases into the atmosphere creating 'a runaway warming effect'. More startling, perhaps, is the report's assertion that, '[d]ue to past, present and near-future greenhouse gas emissions and heat stored in the [Arctic] ocean' (Schoolmeester et al. 2019, p. 13), these temperature rises are locked in until at least the mid-twenty-first century, no matter what action is taken in the present.

How then, as Bruno Latour asks (2014, p. 1), are we supposed to react to such realisations? Greater civil disobedience is certainly one response. So too is a response structured around a blind faith commitment to sustaining things as they are, one only has to consider the popularity of the 'Keep Calm and Carry On' slogan in austerity Britain to understand the psychological appeal such an approach may have. There is the temptation of technological salvation in vast projects of geoengineering as well as the pursuit of 'lifeboat capitalism', where ecological survival is predicated on a discrimination rationalised by property rights and ethnic supremacy (Mitropoulos 2018). Yet, in a situation where the British commentator Richard Seymour (2019) can write in a national newspaper that the end of the world is 'the most plausible scenario', we also need to begin to engage with questions of profound loss concerning our relationship to ourselves, each other, and the planet such relationships are dependent upon.

That loss is a central condition of contemporary experiences of climate change seems self-evident. However, to approach 'the thinkability of disaster' (Braidotti 2002, p. 1987), that is to understand ourselves as being situated within an apocalyptic moment and what action may be

DOI: 10.4324/9781003128854-7

forthcoming from this realisation, there also has to be a consideration of mourning. For Sigmund Freud, in his foundational study of mourning and melancholia, mourning is a reaction to 'the loss of a beloved person or an abstraction taking the place of the person' (2006, p. 310). Importantly, Freud saw mourning as a time-limited process where closure is the goal and any interference with the process of mourning is deemed to be 'pointless, or even damaging' (2006, p. 310). To respond to ideas of mourning and loss in the contemporary ecological moment requires a step away from this Freudian model and to observe the way that loss 'has made a tenuous "we" of us all' (Butler 2004, p. 20). This invocation of mourning's relational and communal potential has clear ecological resonances as it speaks to the way particular practices of mourning can be used to more clearly tether diverse communities of loss, both human and nonhuman, which operate at dissimilar scales, to a common project. Mourning, therefore, is a 'unifying work' (Willox 2012, p. 142). It allows us to more clearly mark the ways in which human life is deeply integrated not only with the flourishing of nonhuman activity but also with its degradation and harm. This sense of what Robin Mackay has called 'geotrauma', the proposition that '[t]rauma is not personal, and the time of the earth is recorded, accreted, knotted up inside us' (2012), can be used to mourn the loss of certain forms of material practice and conceptual relationships we hold towards the earth. Here, Freud's pointedly human and individualistic study of grief becomes entangled with an unaccountable number of nonhuman loci, while the mechanisms for the generation of loss and mourning are uprooted from the pathologies of the individual subject to the communal relations of societies, species, and biota.

The writings of the American poet Jorie Graham can offer a productive account of this more generous and ecologically minded conception of mourning. A Pulitzer Prize winning poet, Graham's poetry has paid close attention to the relationships between language, time, and perception since the publication of her first collection in 1980. More recently, her work has become increasingly concerned with how these phenomena have been altered by the ongoing climate crisis. Where earlier collections like 2002's *Never* and 2008's *Sea Change* are an account of a society moving towards a series of environmental tipping points, Graham (2003) calling these poems an attempt to rebuild 'the shattered community of the *we*', 2017's *Fast* is Graham's first collection to explicitly engage with the reality that the end of the world may be inexorably approaching.

Literature is no stranger to such apocalyptic imaginings and can play a prominent role in the way society tackles the threat of extinction, offering a zone in which the difficulties of signifying death and the ethical implications associated with such signification can be played out. The poem, in particular, is an encounter with a real event which is inexpressible, 'yet demands to be put into words' (Watkin 2004, p. 2). This embodied aspect of writing, the way the shape and lineation of a poem can recall or stand in for the mourned object, represents a crucial aspect of mourning itself: it is not just an abstract process but always a condition of corporeality. It is 'not something from which we can escape' (Willox 2012, p. 148). This encompassing nature of mourning is central to Jacques Derrida's response to Freud's question (2006, p. 311) on the work of mourning. For Derrida, mourning becomes not only the project of life (to provide the proper ethical grounds for the consideration of others) but also the grounds for political action more generally, arguing that there is 'no politics without an organization of the time and space of mourning' (Derrida 1993, p. 61). As such, to consider Graham's response to mourning and its relationship with the ecological condition of the planet is to observe the ways in which the language of mourning can be at the forefront of any new political consciousness which emerges from the work of mourning itself.

Speed as the condition for mourning

To engage with the full discursive possibilities of mourning, its ethical and political capacity to '[bind] us together, as friends, lovers, or even – as troubled a concept as it is – communities' (Engle 2007, p. 62), there is a need to account for a variety of different elements in *Fast*. As such, this chapter will consider what Graham is mourning and how, the idea of geotrauma – a word Graham uses only once in the collection's final poem, but which nevertheless casts a large shadow over her writing as a psychological response to the destruction of the planet, and the force of poetic articulation as a potential site of new political subjectivities in a world which seems to be quickly passing us by. First, however, there needs to be a consideration of the experiences of speed and acceleration, as suggested by the collection's title, as the condition for mourning in these poems.

The experience of time is becoming increasingly vexed in the Anthropocene. The geographer David Matless (2018) has suggested the term Anthroposcenic to register the ways in which we can encounter the experiential convergence of past and present landscapes. This complex temporality builds on Caitlin DeSilvey's (2012, p. 34) work on 'anticipatory history'. This is a methodological approach which seeks to understand the ways in which the Anthropocene gives time back to landscapes which have been framed as immutable and unchanging. While these critical frames emphasise the way that the past seems to ghost the Anthropocene present, of increasing importance is how experiences of the future are registered in our contemporary moment. Looking forward, the design critic Tony Fry discusses a communal experience of 'defuturing', wherein the Anthropocene distorts our understandings of futurity and agency, our unsustainable practices meaning 'we are now at a point when it can no longer be assumed that we, *en masse*, have a future' (Fry 2009, quoted in Pinkus 2013, p. 195). Similarly, Rosemary Randall (2009, p. 119) has written on the psychological impact of climate change, finding that even if the present continues to feel safe, it does so 'at the expense of the future becoming terrifying'.

Alongside this sense of an acceleration towards a future which seems all encompassing in its apocalyptic tenor are experiences of technological and social change which 'outrace comprehension' (Morris 2006, p. 146) and which in Paul Virilio's words (2000, pp. 115–116) 'blast away all delay, all duration'. Even as Graham sets out the stakes of what it is that is being lost in *Fast*, she is attentive to the way these pressures of time and tense play out at the level of embodied psychological response. In the poem 'Reading to My Father', the focus of the speaker slips across a fuzzily held image of the organic body of the ill father interfaced with machinery and a virtual landscape. The presence of this interaction between the organic body and inorganic simulation of the external environment delineates an anxiety in these poems that even death is becoming denatured, somehow disembodied. For Graham, therefore, the emphasis is on how such grieving at the alienation of the body from its immediate environment can go beyond the private expression of loss to encompass a fully realised mourning at a planetary scale.

Speeding towards the time of final arrival, whether that be death of the individual or systemcide itself, haunts these poems. It is the experiential edge which the rhythm and pulse of the language often foresees itself coming up against, a sense of foreclosure which troubles the poem's linguistically generative power. In the eponymous title poem 'Fast', speed and acceleration become tangled with capacities for articulation and self-definition. Starting with an association between speed and activation or access, Graham seems to have a fruitful time elaborating on different kinds of loads: upload, download, to carry a load, lode as a rich mineral source. Yet, these repetitions do not seem to generate any discursive or semantic momentum in the poem. They allude to a general field of meaning (cybernetics and software) but are themselves deployed in a way which is too fragmented and turbulent to cohere. If one experience of speed and acceleration in these poems is as the context to experiences of extinction, as well as speed in terms of the acceleration past, and accruing of, tipping points and traumatic ecological events (storms, tsunamis, flooding, and so on), then the other is 'an image of dynamism' promoted by 'the acceleration of financialization, driven by the new computing and cybernetic technologies' (Noys 2014, p. 10). This almost illusory sense of speed, or perhaps more properly an idea of speed for speed's sake, the fetishising of speed as the proper condition of modern life, is at play in these odd non-generative linguistic repetitions. The valorisation of speed as a way of taking in information without processing or absorbing it is also present; the use of dashes in 'Fast' hurries the reader on, encouraging them to jump from partial phrase to incomplete sentence, at the same time, as they fracture and distort conventional grammar, making it harder to hold onto the present moment of the poem, a space in which the play of Graham's language can be evaluated and engaged with on its own terms.

Geotrauma and the work of mourning

Arrival, shape, and meaning, words Graham uses in the poem 'Incarnation' to summon a desired for enlargement of bodily presence in the 'now' of the contemporary moment, help to emphasise how Graham's conception of political and social life is grounded in concerns over duration and embodiment. As Helen Vendler argues (1995a, p. 85), there is an impulse towards 'total coverage' in Graham's writing, whereby the 'human body is replicating *itself* in its aesthetic body of words'. Graham herself notes a desire for a kind of poetry which frustrates abstract and metaphorical readings, instead preferring a form of writing which compels 'the awakening of the rest of the reading sensibility – intuition, the body' (Gardner 2006). In the ecological context in which this chapter's reading of Graham is situated, this means that there is no gap between text and world. It is through the 'earthly, terrain-orientated lateral search' of her verse (Vendler 1995a, p. 79), captured in the strong, lengthy lines of Graham's poetics, that Graham seeks a sustainable relationship between self and world. The specific practices of imaginative writing, acts of description and the arranging of syntax, is what allows Graham to, in her own words, gain 'exposure to the actual physical matter' of place and planet, to have the attentive mind 'imprinted, stained by the world' (Graham 2003).

The poems in *Fast* recognise this desired capacity for entering the world but mourn the seeming unsustainability of this ethical and political practice. Political for Graham because language's ability to absorb and encode the perceiving mind in the world is a way to 'summon presence … in particular it's the presence of others' (Graham 2003). In 'Ashes', Earth is described as worn away, eroded by repeated attempts to be accessed (both by the poet and wider processes of economic extraction); the poet too described as something haunted and ghostly, the spectre of the implausibility of planetary flourishing hanging doomily over her. In 'Self Portrait at Three Degrees' (the title of the poem suggesting something about how climate change warps subjectivity), Graham balances presence with absence, craft with ritual, and the insistence of the present with the confused temporalities of the Anthropocene. The poem is dense and slippery, moving between voices, tenses, and desires without ever inhabiting one position for a sustained amount of time as it makes its way towards a consideration of the concept of dwelling. The poem seems to cross-examine itself, posing questions which go unanswered or having a conversation in which a crucial interlocutor remains obscured to the reader. A recurrent concern of these poems, the question of voice, who is speaking and from what subject position, makes it difficult to gauge how 'Self Portrait …' approaches and understands this idea of dwelling. While seemingly assertive, the final injunction to dwell, which repeats throughout the poem and on which it ends, shorn of any elaboration and placed on the uncertain ground of the preceding lines, rings oddly hollow. The insistence of its earlier uses does not seem to sufficiently counter the sense the speaker has of their own lack of materiality, their sketchy, soluble corporeality. Similarly, the forceful isolation of that last use of dwell exists outside the chaotic swirling of voice and tense in the preceding lines, unable to stand up to the speed and turbulence of the contemporary moment, the poem leaving undefined those central terms of anthropos and human.

Elisabeth Ellsworth and Jamie Kruse (2012, p. 6) have argued that climate change is 'making geologic realities sense-able with new physical intensity and from new angles of thought as a situation that we live within, not simply as something "out there" that we study'. This question of physical intensity, of climate change as something we live in, has also been recorded as an expression of psychological distress. An article in the journal *Australasian Psychology* outlined solastalgia as 'the lived experience of the physical desolation of home' (Albrecht et al. 2007, p. 96). While this articulation suggests a smaller scale relationship, between the self and one's immediate environs, for Graham the mourning of dwelling permits a recognition of the relationship between intimate bodily encounters and planetary phenomena. Returning to the opening poem of the collection 'Ashes', we can see how Graham attempts to think through the intertwining of body and planet, human and nonhuman to account for these larger zones of mourning. In the small verbal and graphical slip in the opening lines between plant and planet, the re-positioning of a missing vowel, Graham radically expands both her own relation to the external world and permits a vast array of processes and phenomena into the bounds of mourning. To see human and planet as one body, a continuation of shared processes in which sensory and geological experiences blur into and across each other is to attend to the way an ecologically minded mourning can open us up to a wide arrange of vulnerabilities, to make a claim for the profound material connection between differently constituted bodies. It is this alignment and collapsing into each other of bodily experiences which makes possible for Graham an ethical response to climate change. It is also this arrangement of scales of lively bodies (watery, mineral, fleshy) which is threatened by the dematerialisation of contemporary life, its speeds and vectors pressing down on the bodily time of presence and attention.

This level of association between self and planet, body and place, sees Graham's writing echo Marx's observation (1970, p. 112) that 'man's physical and spiritual life is linked to nature' in such a profound way that 'nature is his *body*, with which he must remain in continuous interchange if he is not to die'. In mourning a planet threatened by climate change, Graham is also mourning the loss of something in us which is intimately bound up with planetary life. For Freud (2006, p. 311), the work of mourning is centred on the 'loss of a loved one' or the 'love-object', yet this singular focus on the isolated object seems inadequate for the kinds of mourning climate change demands of us. We are compelled to mourn for entire species, and for futures that now seem beyond us at such an intense scale that it raises questions about the adequacy of the tools we have to mourn with.

A poem like 'From Inside the MRI', for example, articulates how the same sensory yearnings for larger collective entanglements can be overloaded by the demands of technological modernity. The oneness that Graham tries to summon at the end of 'Ashes' has become punctured and perforated by the body's placement within the medical machinery of diagnosis, the solidity of the human body described as its own site of excavation by medical science, clearly linking the logics of a technologically advanced medical science to process of industrial extraction. Later, the drips and veins of the medical procedure re-emerge as the xylem tubes of plant life, the poet straining to listen to microscopic processes of water collection and photosynthesis from her location inside the MRI machine. However, the urge to listen to water evaporating from the leaves of plants becomes muddled with a news report playing somewhere else in the hospital, detailing the brutalisation of Yazidi people by Islamic State in Syria and Iraq. Here the pressure of time remerges, the endless piling up of minutes upon minutes which squeeze out duration, the arrows which link these phrases becoming less the assertive association of bodily sense with vegetable life and more the unstoppable rush of our extended moment towards war and devastation. Crucially, it is the weighing up and accretion of moments and sensation without any capacity for active engagement or reshaping, an inability to disaggregate daily sensory experiences of violence and trauma from what is valuable or needs to be sustained.

While Graham never fully embraces the ecological nihilism which is present in these poems, she does confront the possibility that the ability to properly mourn the losses of the natural world, to mourn our capacities to interface and engage with material life, are themselves under pressure from the speed and velocity of climate change. For Julia Kristeva, both practising psychologist and literary scholar, the work of mourning centres on an understanding of loss, with depression or melancholia being the inability of 'knowing how to lose' which entails 'a loss of my being' (Kristeva 1989, p. 5). To be unable to mourn is a threat to the coherence of self and it is notable how acceleration and speed in these poems often come across, despite its physical pressures on body and language, as an almost airy phenomenon for Graham. Accelerationism is not only seemingly disembodied itself, but speed and velocity in these poems exert force on and destabilise the differently constituted bodies in these poems. A kind of ghostly sorrow which makes mourning itself difficult and seeks to unbody the world, disintegrating the grounds for mourning. Without remains, without some kind of object to orientate examinations and sensations of loss around, the wider work of mourning, its political and ethical dimensions, becomes hard to sustain. For Derrida mourning

> consists always in attempting to ontologize remains, to make them present, in the first place by *identifying* the bodily remains and by *localizing* the dead… Nothing could be worse, for the work of mourning, than

> confusion or doubt: one *has to know* who is buried where—and *it is necessary* (to know—to make certain) that, in what remains of him, *he remains there*. Let him stay there and move no more!
>
> (1994, p. 9)

While important qualifications need to be made here, particularly in the current ecological moment where what is mourned is both particular species which can be localised and identified and wider planetary processes which encompass and make possible the work of mourning in the first place, the emphasis on the potency of remains and matter, of the material stuff of mourning, is clear. Without some sense of remains, there can be no real closure, and without closure mourning seeps into melancholy and depression.

It is not just the matter and location of mourning which is coming under threat in these poems but also its temporality. For mourning to operate as an ethical and political process in Derrida's model, it relies on an understanding of time in which one must have already 'marked in the present what would be repeatable into a future' (Colebrook 2017, p. 119). This ethical present is important for Graham and is differentiated from an acceleration of the now, the current moment imposing itself 'on consciousness as a moment in extended crisis, with one happening piling on another' (Berlant 2011, p. 7), because it allows for duration and for tracing the relations that have brought the present into existence:

> all this living in the present moment, this limited terrain left us to quest in as the slower temporality journeying requires is eliminated. All this narrowing to a *now* in which there's only room for *effect*, not enough room for *cause*, if you will—and so no duration in which to experience personal accountability. There's precious little room.
>
> (Graham 2003)

In 'Self Portrait at Three Degrees' the speaker is unsure of where they are situated in the flow of time, asking if they are ahead or behind the normative linear progress of temporality. The speaker in 'Post-Human', grieving the loss of an intimate relation, finds themselves unsure about what exactly there is left to mourn and the timeframe in which the body remains to be mourned in, asking if it has all gone at once or piece by piece. Again and again, these poems return to speed not just as a condition of mourning but as a phenomenon which places immense pressure on the sustainability of mourning as an encompassing practice of attention and witnessing.

If Freud predicates mourning, in opposition to melancholia, as a process of closure in which the object of loss is recognised and its psychic function replaced, Graham's poems register both the strange temporalities of the Anthropocene which make closure difficult and the collapse of assumptions about a stable, material space in which the performance of mourning is able to take place. Importantly, however, they push beyond the bounded, ego-focused notion of Freud's psychoanalysis to offer a language to mourn the differently constituted bodies of human and planet. Even if these discourses are fraught and their sustainability is questioned, it is here that Graham's struggle to carry out the work of mourning is most vibrantly felt.

Political subjectivity at the end of the world

Against the emergence of a world without bodies: all flow, all speed, wholly virtualised, Graham posits small interactions, of being at hand which is also a holding in and between hands, as sites of potential resistance, as a way of staying in the good minute of the collection's epigraph from Robert Browning's 'Two in the Campagna'. This attention to those things which are at hand, close to us and immediate, implicitly less mediated by what Graham views as the interfaces of contemporary technological culture, insists on the necessary entanglement of a politically motivated ethics of embodied experience. In the poem 'Incarnation', the emphasis on tangible things is also a way to have a say in things, the poem playing with the slippage between these different types of thing-ness. Through this, Graham reinforces the idea that it is at the level of intimate bodily encounters that a potential politics for rebuilding communal and ethical capacities can emerge (to have a stake in things here is not just a political but worldly activity). At the same time, Graham does not elevate these small gaps of political and ethical light above the anxieties and pessimism this chapter has also delineated. The light of the day will fade, the body (both planetary and human) will ebb away, cannot be sustained indefinitely, a poem like 'Vigil' presenting us with an image of joy and death bound together, pierced by (and piercing) each other. This is, therefore, a politics and poetics of salvage and remains. It does not seek to avert or overcome the seemingly inevitable (and already ongoing) impact of climate change. Rather, these moments begin to look towards the fraught question of what might come after the end of the world. Douglas Crimp, in arguing for the necessary coupling of 'mourning *and* militancy' in the context of the AIDS crisis in the mid-to-late 1980s, states that such an outcome, a communal and politically charged articulation of loss and grief, can only come from recognising the psychic and emotional violence from which the conditions of mourning spring most readily from:

> if we understand that violence is able to reap its horrible rewards through the very psychic mechanisms that make us part of this society, then we may also be able to recognize—along with our rage—our terror, our guilt, and our profound sadness. Militancy, of course, then, but mourning too.
>
> (Crimp 1989, p. 18)

By refusing to wish away the realities of climate change, and in acknowledging how any sense of political futurity is likely to take place in the aftermath of ecological devastation, Graham begins to confront the realisation that 'the world hasn't ended, but the world as we know it has—even if we don't know it yet' (McKibben 2011, pp. 2–3).

Endings are important to Graham. They provide 'a sense-of-shapeliness', specifically in how 'the formal elements that occur along the way are only audible – or believable – in relation to their extinction, the closure' (Gardner 2006). Endings, of poems, books, life, give meaning and significance to the experiences we group under these nouns. As Helen Vendler has remarked, 'Graham attempts to be in the flow before demarcating it, though she also ponders demarcation' (1995b, p. 244). In the final poem to *Fast*, 'Mother's Hands Drawing Me', Graham tires to negotiate the various mediations of experience the poem is filtered through (its own writing, the mother's drawn representation of Graham) to stay in the flow of the moment, to extend that 'good minute' (Browning 2009, p. 283). It is the desire for the poem not to be an account of her writing of the event of her mother drawing her but to, in some sense, be the event itself, to keep separate the work and craft of the mother from Graham's own poetic practice. Despite this, the looked for moment of pure embodied relationality, of hand to pencil to eye to face, is mingled in its own way with the recognition of a variety of endings which, while extended through the duration of Graham's own writing of this moment, are not stalled forever and force Graham's own commemoration of her mother into the writing of the poem.

For Derrida, when one works at the work of mourning 'one is already, yes, already, *doing* such work, enduring this work of mourning from the very start, letting it work within oneself, and thus authorizing oneself to do it' (1996, p. 172). This encompassing sense of the work of mourning, an understanding that the finitude of life can give it purpose and meaning, permits into the fold of the labour of grief a wide range of tactics which are not conventionally understood as modes of mourning. Successful mourning from this point of view is not the Freudian replacement of the lost object of grief but rather an ongoing process where we 'recognizing that all we have to give to mourning, to the dead, to what we have lost, is in our own living and our own acts of mourning and remembering' (Cunsolo and Landman 2017, p. 10). The real work of mourning, therefore, emerges in the ongoing interchanges between the living and the dead. For Graham this takes the specific relationship of craft and form; the self-proclaimed rushings and scribblings of the poems act as a way of drawing the unseen and unremarked into the open, forcing the poet not just to mourn for what is gone but confront what remains (even if barely) and begin to ask for what purpose can these remains be put.

In Graham's poetry this task of making visible, which stands as the final, resonant word of the collection, small yet defiant, is not an a priori function of poetic language but is something which emerges from specific formal interventions. As such, a useful distinction can be made between the mother's scribblings in the collection's final poem and the depictions of the Charles

River in the poem 'Medium', which evocatively seems to map onto Graham's own writerly activity. In 'Medium', the poetic line channels, scribbles, and erases itself before heading on again, meandering across the page only to accelerate forward, capturing the rhythm, pulse, and visual arrangement of the river as it makes its way through the New England landscape, the lines even turning back on themselves at the right-hand margin as the river meanders through its course. Not only is language here embodied by and embodies the river, but it is also entangled in that wider consideration of body and world, death and process, the river mist rising like a ghostly ectoplasm as if it were the very remains the poet wanted to consider, at once both personal and ecological. Graham pegs this scribbling and re-inscription to processes of mourning and remembrance, of making the unsaid sayable.

Yet, importantly, there seems to be a qualitative difference between the scribbling Graham associates with the river, and by extension her own writing, and the scribbling the mother performs, which seems to transform Graham's rendering of events. In 'Medium', this scribbling is trapped in the same kind of non-generative forms as the language in 'Fast'. The river is depicted as both swallowing itself and spewing itself onto itself, a strange accretive image in which the river is only able to replicate a copy if itself (suggesting something of Graham's wider anxieties about virtual life denaturing the external world), and a noted juxtaposition to Heraclitus's famous assertion that one can never step into the same river twice. This sense of stasis also emerges through Graham's depiction of the river's flow, even though the river is carried forward it does so only by drifting, lacking purpose and direction, as if the river will simply dissipate like so many other bodies in these poems, rather than find its way to the sea.

In 'Mother's Hands', this passiveness is replaced by a sense of active change and becoming. The injunction to draw (here said to the mother by the poet) becomes extended to the poet asking how they are drawn, that is rendered both within the pages of the poem and equally in the world through the attention of others, offering the beginnings of a reaction to the spectrally withered and diminished notion of inhabitation staged in the opening poem, 'Ashes'. The speaker is not only rendered by the movements of her mother's hands but, in the process of witnessing this materialisation of themselves through the gestures and movements of the dying woman's hands, find themselves drawn into the world, and drawn into the word-body of their own writing. This drawing out and into the world is located here not as the private activity of grief but as a communal working through and recognition of both the work of the dead and the labour of mourning the living can carry out. The speaker in this poem, and by extension the reader, is made to see themselves as the dead or dying see them. It is only in this watching and being watched in the act of marking and re-marking that the light in 'Vigil', which was so quickly passing, threatening to become substanceless like the body being mourned, can be given mass, can be said to recur into a durable and ethically charged present.

'When loss remains unspoken, neither grieved nor worked through, then change and adjustment cannot follow' (Randall 2009, p. 119). At a primary level this could be said to be the main achievements of mourning. In allowing for 'change and adjustment' however, it is right to ponder who is articulating these experiences of loss and to what ends change is being fostered. This chapter, in examining Graham's attempts to pay attention to the passing of certain configurations of worldly encounters, has marked out the way the earth seems increasingly untenable as a site for dwelling and as an interface for an embodied presence which is at its heart a political and ecological process. It has also paid attention to what Graham does with this recognition of loss, with the remains of these ways of being in the world, which find themselves threaded through anxieties of speed, acceleration, and dematerialisation. For William Watkin's, the work of the literary elegist or memorialist is to 'create a physical location for remembrance, [free] from the manipulation of the material world' (2004, p. 7). Despite the recurrent fears over being bodiless, the physical arrival of *Fast* as a collection of poems into the world acts as this 'physical location' for remembrance. Yet, it only does so because the poems themselves engage the possibilities of mourning in a way which is resistant to the static connotations of memorialisation as a practice. Rather than trying to exert mastery over inert remains, Graham's writing here is a reinvigoration of liveliness in an extended and collective conception against the deathliness of both contemporary capitalist forms of production and the limitations of the particular singularity of the poet's corporeal existence.

References

Albrecht, G., Sartore, G., Connor, L., Higginbotham, N., Freeman, S., Kelly, B., Stain, H., Tonna, A., and Pollard, G. (2007) Solastalgia: The Distress Caused by Environmental Change. *Australasian Psychology*. 15, 95–98.

Berlant, L. (2011) *Cruel Optimism*. London: Duke University Press.

Braidotti, R. (2002) *Metamorphosis: Towards a Materialist Theory of Becoming*. Cambridge: Polity.

Browning, R. (2009) *The Major Works*. First published 2005. Oxford: Oxford University Press.

Butler, J. (2004) *Precarious Life: The Powers of Mourning and Violence*. London: Verso.

Colebrook, C. (2017) The Twilight of the Anthropocene. In: A. Johns-Putra, J. Parham and L. Squire, ed. *Literature and Sustainability*. Manchester: Manchester University Press, pp. 115–136.

Crimp, D. (1989) Mourning and Militancy. *October*. 51, 3–18.

Cunsolo, A., and Landman, K., (2017) Introduction: To Mourn Beyond the Human. In: A. Cunsolo and K. Landman, ed. *Hope at the Heart of Ecological Loss and Grief*. Kingston: McGill-Queen's University Press, pp. 3–26.

Derrida, J. (1993) *Aporias*. Translated by Thomas Dutoit. Stanford: Stanford University Press.

Derrida, J. (1994) *Specters of Marx: The State of the Debt, the Work of Mourning, and the New International.* Translated by Peggy Kamuf. London: Routledge.
Derrida, J. (1996) By Force of Mourning. Translated by Pascale-Anne Brault and Michael Naas. *Critical Inquiry.* 22(2), 171–192.
DeSilvey, C. (2012) Making Sense of Transience: An Anticipatory History. *Cultural Geographies.* 19(1), 31–54.
Ellsworth, E., and Kruse, J. (2012) Introduction. In: E. Ellsworth and J. Kruse, ed. *Making the Geologic Now: Responses to Material Conditions of Contemporary Life.* New York: Punctum Books, pp. 5–26.
Engle, K. (2007) Putting Mourning to Work: Making Sense of 9/11. *Theory, Culture & Society.* 24(1), 61–88.
Freud, S. (2006) Mourning and Melancholia. In: A. Phillips, ed. *The Penguin Freud Reader.* London: Penguin, pp. 310–326.
Gardner, T. (2006) Interview with Jorie Graham [online]. *Jorie Graham.* [Accessed March 23 2020] Available www.joriegraham.com/gardner_interview_2006
Graham, J. (2003) The Art of Poetry No. 85 [online]. *The Paris Review.* 165. [Viewed 23 March 2020]. Available www.theparisreview.org/interviews/263/jorie-graham-the-art-of-poetry-no-85-jorie-graham
Graham, J. (2017) *Fast.* Manchester: Carcanet.
Harvey, F. (2019) Sharp Rise in Arctic Temperatures Now Inevitable – UN [online]. *The Guardian.* 13 March 2019. [Accessed 7 May 2020]. Available www.theguardian.com/environment/2019/mar/13/arctic-temperature-rises-must-be-urgently-tackled-warns-un
Kristeva, J. (1989) *Black Sun: Depression and Melancholia.* Translated by Leon S. Roudiez. New York: Columbia University Press.
Latour, B. (2014) Agency at the Time of the Anthropocene. *New Literary History.* 45(1), 1–18.
Mackay, R. (2012) A Brief History of Geotrauma [online]. [Accessed 7 May 2020]. Available http://readthis.wtf/writing/a-brief-history-of-geotrauma/
Marx, K. (1970) In: D. Struik, ed. *Economic and Philosophic Manuscripts of 1844.* Translated by Martin Milligan. London: Lawrence and Wishart.
Matless, D. (2018) The Anthroposcenic: Landscape in the Anthroposcene [online]. *British Art Studies.* 10. [Accessed 23 March 2020]. Available www.britishartstudies.ac.uk/issues/issue-index/issue-10/landscape-anthroposcene
McKibben, B. (2011) *Eaarth: Making a Life on a Tough New Planet.* New York: Times Books.
Mitropoulos, A. (2018) Lifeboat Capitalism, Catastrophism, Borders [online]. *Dispatches.* 1. [Accessed 20 July 2020]. Available http://dispatchesjournal.org/articles/162/
Morris, A. (2006) The Act of the Mind: Thought Experiments in the Poetry of Jorie Graham and Leslie Scalapino. In: Robert Crawford, ed. *Contemporary Poetry and Contemporary Science.* Oxford: Oxford University Press, pp. 146–166.
Noys, B. (2014) *Malign Velocities: Accelerationism and Capitalism.* London: Zero Books.
Pinkus, K. (2013) Thinking Diverse Futures from a Carbon Present. *Symploke.* 21(1–2), 195–206.
Randall, R. (2009) Loss and Climate Change: The Cost of Parallel Narratives. *Ecopsychology.* 1(3), 118–129.
Schoolmeester, T., Gjerdi, H. L., Crump, J., Alfthan, B., Fabres, J., Johnsen, K., Puikkonen, L., Kurvits, T., and Baker, E. (2019) *Global Linkages: A Graphic Look*

at the Changing Arctic [online]. UN Environment and GRID-Arendal. [Accessed 7 May 2020]. Available www.grida.no/publications/431

Seymour, R. (2019) The Climate Strike Is a Source for Hope – But New Research Shows It Might Be Too Late [online]. *The Independent*. 16 March 2019. [Accessed 7 May 2020]. Available www.independent.co.uk/voices/climate-strike-protests-arctic-melting-a8825936.html

Vendler, H. (1995a) *The Breaking of Style*. London: Harvard University Press.

Vendler, H. (1995b) *Soul Says: On Recent Poetry*. London: Harvard University Press.

Virilio, P. (2000) *The Information Bomb*. Translated by Chris Turner. London: Verso.

Watkin, W. (2004) *On Mourning: Theories of Loss in Modern Literature*. Edinburgh: Edinburgh University Press.

Willox, A. C. (2012) Climate Change as the Work of Mourning. *Ethics and the Environment*. 17(2), 137–164.

7 Meaningful life at the end of times

Ageism and the duty-to-die in *Logan's Run*

James A. Tyner

> Why had she believed in Sanctuary? Insane. Impossible. Why hadn't she been like all the others who had accepted Sleep?
>
> Now the dark figure, in black, stood over her, but she did not look up. She didn't beg because begging was useless.
>
> She saw the Gun come up.
>
> There was a soft explosion.
>
> That was the last thing she heard.
>
> And the last thing she felt was raw, blinding agony, as the homer struck, burned, ripped and unraveled her.
>
> (Nolan and Johnson 1967, 2–3)

So begins William Nolan and George Johnson's (1967) post-Apocalyptic and dystopian classic *Logan's Run*.[1] The setting of their novel is the twenty-third-century United States. The premise, as with so many works of science fiction, plays on the fears of eugenics and state authority. Indeed, very much of a product of its time, *Logan's Run* addresses the specters of overpopulation, food insecurity, and nuclear war. *Logan's Run*, however, provides also an opportunity for reflection, that is, an opening into the societal valuation of life itself and, especially, the dangers of state-enforced euthanasia directed against the elderly (Snaith 2017; Ahn 2020).

In recent years, geographers and other social scientists have focused attention on the politics of life and death. This work, in particular, has emphasized both the *biopolitics* and the *necropolitics* that condition particular interventions that together facilitate or hinder one's survivability (McIntyre and Nast 2011; Davis, Isakjee, and Dhesi 2017; Laurie and Shaw 2018; Tyner 2019). In turn, scholars have rearticulated an 'ethos of living' and an 'ethics of care,' namely, an engagement with, and an attentiveness to, that which makes life both possible and meaningful (Lawson 2007; Till 2012; Emmerson 2019). As Emma Laurie and Ian Shaw (2018: 8) write, "Violent conditions burn in the background of daily life." They conclude: "Violent conditions forcefully constrain, traumatize, and poison the very resources of our becoming. This is an injustice lodged in the fibers and skeletons of being" (p. 15).

DOI: 10.4324/9781003128854-8

A related strand of work focuses on geographic imaginaries and how fictional accounts of apocalyptic futures offer potential sites in which "the politically possible is constituted" (Schlosser 2015, 309; see also Dittmer and Spears 2009; Swyngedouw 2012; Bettini 2013). "Expectations of a dystopian or apocalyptic future," Florian Mussgnug (2012, 333) writes, "are central to twentieth and twenty-first century culture, and constitute a pervasive influence on popular culture, religious identity, and political ideology." Christianity, for example, has created a sacred history—an account that binds together the moments of life into a continuous narrative whereby those who live a righteous life "may pass the test of time—and enjoy life eternal" (Fenn 2003, 108). In other words, dire predictions about the 'end of time' condition one's duty to live a particular, meaningful life. Indeed, prescriptions on how to live are fundamental to apocalyptic narratives. As Pitetti (2017) explains, although the term 'apocalypse' often connotes 'disaster' or 'cataclysm,' its derivation from the Greek language actually refers to an 'uncovering' and is expressed most vividly in the Bible's 'Book of Revelation.' In this sense, apocalyptic narratives provide insight into *meaning*, specifically, *meaningful life* and are thus revelatory not so much in the scientific details of catastrophic events but instead as a revelation of what societies deem important or valuable. More precisely, faced with an impending or actual cataclysm, how are humans (and non-human animals) valued? Who, in other words, has the right to live or the duty to die?

Apocalyptic science fiction, especially, provides a lens through which we can re-evaluate the potential violent conditions of a post-Apocalyptic world and the promise or failure to *care*. As Connor Pitetti (2017) explains, apocalyptic literatures generally fall within two forms. On the one hand, cataclysmic events—nuclear Armageddon, climate change, famine, and pandemics—are used to give structure to societal transformations, that is, 'end of time' events are presented as ruptures or pivots, thereby enabling readers to contemplate the valuation of life *before* and *after* the event. On the other hand, catastrophic occurrences are presented as slowly unfolding over time, thus allowing readers to consider the continuities of interventions that augment or dissipate the crisis. *Logan's Run* falls under the rubric of the latter, whereby a series of cascading and catastrophic events segue into ever-more dire situations and, through the course of the novel, readers witness the incremental yet inexorable slide into a dystopic descent of corporeal callousness. In *Logan's Run*, the 'end-of-time' scenario portends a society whereby life has little meaning beyond the mere passing of time; and those who simply want to live 'one more day' are hunted with deadly efficiency. As such, in this chapter I argue that *Logan's Run* articulates an ageist future whereby draconian practices that otherwise would seem inconceivable become normalized.[2]

Critics of the novel suggest that *Logan's Run* is anti-youth; that the novel portrays youth in a negative light and thus speaks to the generational gap of the late 1960s. My read is somewhat different, in that I view *Logan's Run* as a critique against *ageism* in American society. Furthermore, I forward the idea that *Logan's Run* aptly foreshadows the valuation of life and death within neoliberalism and the security state. For the novel presents an exaggerated account of eliminating those who are perceived to be no longer productive, in this case, the elderly and unfit. Young adults *per se* are not being killed; rather, they are prevented from aging—from becoming non-productive members of society. Specifically, I read in *Logan's Run* a cautionary tale of eugenics, population engineering, and state-authorized euthanasia, whereby one's duty-to-die is framed as a necessary obligation to society itself. To this end, *Logan's Run* offers insight into our contemporary moment whereby the aged are expendable and their loss is a necessary feature of a functioning capitalism underscored by a neo-Malthusian specter of overpopulation.

The apocalyptic politics of *Logan's Run*

Logan's Run is set in a post-Apocalyptic world wrought by human fallibility. The 'end' began gradually, with the specter of overpopulation and food insecurity: "At the end of the Twentieth Century … when men spawned like microbes on a culture dish, the great problem was food. The fourth horseman rode the land and his name was Famine" (Nolan and Johnson 1967, 5). Initially—as often is the case—humanity turned to technology and the colonial impulse. "Man reached for the planets," we learn, but found nothing but "puddled gas and frozen stone" (Nolan and Johnson 1967, 5). And having thus abandoned space, attention was directed toward the oceans and the cultivation of aquaculture. Huge artificial communities were constructed and thousands of technicians and their families labored to produce food in vast food-processing centers. These too proved not to be the hoped for technical panaceas they were designed. Undersea earthquakes literally shattered many of the structures with deadly consequences.

Non-technological solutions followed, including the draconian regulation of fertility and mortality. In actuality, these corporeal solutions *always* existed albeit not without opposition. In the late twentieth century, for example, protests over governmental intervention in the management of reproduction began. The catalyst for protest was the 39th Amendment to the Constitution, known as the Compulsory Birth Control Act. Protests were directed predominantly at two governmental bureaucracies: the National Council of Eugenics and the Federal Birth Study Commission. After two weeks of fighting—known as the 'Little War'—the government was overthrown, to be controlled by the youth.

Having achieved a political coup, subsequent debate centered on how to solve the population crisis. Within this context a 16-year-old boy, named Chaney Moon, initiated a fanatical movement encouraging followers to embrace 'Sleep'—a euphemism for death. As detailed in the novel,

> Young America accepted this bold new method of self-control, and the Thinker [a computer] was programed to enforce it. Eventually

> all remaining older citizens were executed and the first of the giant Sleepshops went into full-time operation in Chicago. One thing the young were sure of; they would never again place their fate in the hands of an older generation. The age of government by computer began. A maximum age limit (21-years) was imposed with the new system, and the original 'DS' (Deep Sleep) units were formed. By 2072 all the world was young.
>
> (Nolan and Johnson 1967, 123)[3]

Thus, *Logan's Run* addresses overpopulation and scarce resources; but it also speaks to a perverted form of health-care rationing, a situation where citizens have an obligation to die in order to preserve the state. *Logan's Run* is also a parody of advanced capitalism, of the ideal to promote a youthful, able, productive society.

In *Logan's Run*, all people upon birth have radioactive 'flower' crystals embedded in the palm of their right hand. These crystals decay at a set rate, and change colors to mark the passage of years: sun-yellow to electric-blue to a warm blood-red, the colors of the crystal provide a biological measure of one's allotted time until, upon reaching their 21st birthday, the crystal turns to a dull black. When people reach their mandated end of time, they are to report to a *Sleepshop* where they are painlessly euthanized, a process known as 'Deep Sleep.' To be sure, not all citizens want to embrace 'Deep Sleep'; many, in fact, wish to live just a little bit longer and thus flee, in a vain attempt to reach a mysterious place called 'Sanctuary.' Those who flee are tracked down, hunted, and executed by 'Sandmen,' members of the state security apparatus. The novel focuses on Logan, a Sandman, who attempts to escape his own state-sanctioned death at age 21.

The idea—and ideal—of obligation runs strong throughout the novel. In an early scene, Logan is talking to Sawyer, another Sandman. Sawyer, who is clearly fearful of dying, questions the reasoning behind 'Deep Sleep':

> Don't get me wrong, citizen. I'm no coward. I'm not going to run. I have my pride. The system is right, I know that. World can only support so much life. Got to be a way to keep the population down…. I've been loyal and I won't change now.
>
> (Nolan and Johnson 1967, 5)

Sawyer's statement is interesting and revealing, in that it speaks both to the rightness—the justness—of the system and also to one's duty. Sawyer has internalized an obligation to sacrifice *his life* for the betterment of state and society.

Throughout the novel, such an obligation—a 'duty-to-die'—is continuously reinforced through technology: loudspeakers, view-screens, and so on. In one scene, for example, Logan is witness to a robotic re-enactment of the 1862 Battle of Fredericksburg. A monotonous voice drones overhead, describing the scene for on-lookers: *"Duty," a loudspeaker blared above the din.*

> That's what you'll see here today, citizens. Loyalty. Courage. The willingness to die for one's country in order to preserve it. The Civil War was fought by seventeen- and eighteen-year-olds, men willing to die for their cause. They did not question their duty or flinch from the face of death. They sacrificed themselves willingly, gloriously.
>
> (Nolan and Johnson 1967, 102)

The analogy is clear. Before one becomes too old, or too frail, one has an obligation to die for the state. And if this does not materialize, gloriously, on the battlefield, then one must willing to choose death—with dignity—on the home front.

The contemporary duty-to-die movement

Zygmunt Bauman (2004) writes about 'redundancy' in contemporary society. Such a condition, according to Bauman (2004, 11),

> whispers permanence and hints at the ordinariness of the condition. It names a condition without offering a ready-to-use antonym. It suggests a new shape of current normality and the shape of things that are imminent and bound to stay as they are.

He continues:

> to be 'redundant' means to be supernumerary, unneeded, of no use—whatever the needs and uses are that set the standard of usefulness and indispensability.... To be declared redundant means to have been disposed of because of being disposable—just like the empty and non-refundable plastic bottle or once-used syringe
>
> (p. 12)

This is an important distinction—separate from the concept of 'surplus populations' (Shaw and Waterstone 2019). Surplus populations—the infamous 'reserve army of labor' so crucial for capitalism—are *necessary but contingent*. To be redundant suggests that capitalism no longer has any *productive* use for those who are labeled as such. Redundancy also is an antiseptic word. It is found in the vocabulary of the bureaucrat; of the administrative 'bean-counter' who makes rational calculations to balance the bottom-line. It speaks the language of *measured time*, of a biopolitical accounting of life itself. And much like beans *appear* the same, so too are individuals rendered identical within the bio-arithmetic of capitalism. As Bauman (2004, 12) explains, "routinely, people declared 'redundant' are talked about as mainly a financial problem. They need to be 'provided for'—that is fed, shod and sheltered." Of course, in contemporary society no *individual* is able to live autonomously, completely independent of assistance or input from others. Those who are most wealthiest in society for example are entirely dependent on the labors of others: *someone* must grow, process, and distribute their food; *someone* must ensure an adequate supply of clean water; *someone* must build their homes and resorts; *someone* must provide necessary medical and health care. We are all dependent; we are all entwined in a series of complex social relations (Tyner 2016, 132).

But in the sterile world of the 'free' market, the indifference to use value conceals a preference for a narrowly defined, utilitarian productivity. As Lisa Marie Cacho (2012, 19) explains,

> in the era of American neoliberalism, social value and moral behavior are interpreted through and evaluated on economic terms, and, as a result, capitalist logic and ethics prevail in the social sphere as well as the economic and political realms.

She continues that "in neoliberal ways of knowing, the *value of life* is subjected to an economic analysis and assessed accordingly: How has this person contributed to society? What will he or she accomplish in the future?" (p. 33).

Throughout the late nineteenth and early twentieth centuries, concern grew among the academic community that contemporary society—through charity and welfare programs—were 'tampering' with the laws of nature; that 'unfit' people were being allowed to survive and reproduce; that 'lives not worth living' constituted an unbearable financial burden on national and local budgets; and that those deemed undesirable constituted a criminal and violent threat. In other words, 'positive' duties threatened to 'harm' society at large; required was an end to such charitable actions and a concomitant political will to intervene into the management of life.

These biopolitical valuations of life identified a grievous and imminent social problem and demanded an immediate solution. This was not the stuff of science fiction but rather the perceived reality of capitalist science. Fueled by Darwinian concepts of sexual selection and evolution, competing concepts of genetics, Malthusian perspectives on overpopulation, and environmental determinism, widespread fears of racial and societal degeneration—coupled with the ever-present fear of economic collapse—permeated both popular and scientific venues. These fears were augmented by a growing—and ostensibly scientifically documented—knowledge of demographic changes. In the United States, for example, long-established patterns of immigration were drastically changing. Prior to the 1880s immigration was dominated by northern and western European nations; in the decades surrounding 1900, however, the main source areas shifted to southern and Eastern Europe and, increasingly, Asia. Citing these changes, the labor historian and university professor John Commons, in 1907, warned that "the change is one that should challenge the attention of every citizen" (p. 70).

Science eagerly rose to the challenge and set itself the task of identifying the 'proper' form of governmental interference necessary to rationalize demographic change. What emerged was the eugenics movement. As Sheldon Ekland-Olson (2012, 21) writes, the eugenic argument was attractively simple: would it not be better to have a society enriched by those who are productive, healthy, emotionally stable, and smart, than one stifled by degenerate, feeble-minded, disabled, and criminal citizens? Framed as a concern over differential worth, the eugenics movement effectively combined sexist, racist, classist, and nationalist sentiments.

As both an academic field and a political agenda, there was no singular approach to eugenics. Rather, individual practitioners identified themselves largely into four main groups: mainline eugenicists, racial anthropologists, reform eugenicists, and socialist eugenicists. Indeed, what is striking about eugenics was that it could be, and was, adopted by people spanning the political spectrum. And while significant differences existed, what is most significant is that all retained a central tenet in that it was possible—in fact, necessary—to engage in a calculated management of valued life.

Eugenicists differed also in their application of science. Whereas some favored 'positive' eugenics (i.e., policies to increase the contribution of populations deemed most desirable), others advocated 'negative' eugenics (i.e., policies to decrease the biological contribution of inferior or less valuable populations). By way of illustration, in *Logan's Run*, although the regulation of fertility is widespread and deliberate, the predominant method centers on negative eugenics, that is, the euthanasia of people upon reaching 21 years of age.

How can the state call upon its citizens to sacrifice their lives for the betterment of society? This is a question that figures prominently in studies of the Holocaust and genocide; but it informs also much of our current debates on abortion, euthanasia, and capital punishment (Ekland-Olson 2012; see also Tyner 2012). "We all accept that killing is in general wrong," Jeff McMahan (2002, vii) writes, "but virtually all of us also recognize certain exceptions—that is, we conceded that there can be instances in which killing is permissible." Killing in self-defense, for example, is often ruled as an acceptable, defensible, and hence legal reason for taking another's life; likewise, killing in defense of one's country is also considered to be just. Other forms of killing may be justified—such as physician-assisted suicide—but these are often subject to intense debate. Here, following the apocalyptic future portrayed in *Logan's Run*, I limit by discussion to the practice of euthanasia. Derived from two Greek words, *eu*, meaning 'good,' and *thanatos*, meaning 'death,' euthanasia simply translates as 'good death' and is thus redolent of the euphemisms that pepper *Logan's Run*.

In the novel, on 'Lastday,' people are required to appear at one of the many 'Sleepshops' whereupon they are drugged into a state of euphoria and euthanized via lethal gas. In reality, euthanasia comprises a variety of different forms. Euthanasia, for example, may be active ('killing') or passive ('letting die'). For example, the administration of a lethal injection by a physician constitutes an active form of euthanasia, whereas the removal of life-support systems is a type of passive euthanasia. Euthanasia may also be classified by various *relations* of consent. *Voluntary* euthanasia refers to a situation where a person freely and autonomously requests or consents to being killed or allowed to die. However, the notion of 'free' choice is highly contentious and itself becomes a legal decision. Infants, for example, are considered to lack autonomy and thus unable to make their own decisions; the elderly, likewise, are often seen as 'losing' this condition. Consequently, *non-voluntary* euthanasia refers to those cases where it is not possible for an individual who is killed or allowed to die either to give or to withhold consent, such as infants and 'congenitally severely impaired' persons. Lastly, euthanasia is said to be *involuntary* when an individual who is competent to give or to withhold consent is killed or allowed to die either contrary to his or her expressed will or when his or her consent has not been sought (McMahan 2002, 457). It is interesting that in *Logan's Run* people ostensibly volunteer to be euthanized; however, the existence of sufficient 'runners' and the concomitant necessity of forming an elaborate policing unit of 'Deep Sleep' operatives call to question the 'voluntariness' of the practice. Indeed, it is worth speculating why some people zealously support their prescribed date of execution while others resist.

And herein lies the fundamental politics of the post-Apocalyptic future portrayed in *Logan's Run* and the parallels with contemporary society. Throughout the United States and elsewhere, renditions of a 'duty-to-die' movement have appeared (see for example Hardwig 1997; Corlett 2001; Gunderson 2004; George, Finlay, and Jeffrey 2005; Hardwig 2009; Cholbi 2010; Enck 2010; Golden and Zoanni 2010; and Dore 2011). Elsewhere I argue that the 'duty-to-die' movement underscores a deeper animus—and violence—toward the 'dis-abled' and the 'elderly': two figures that are progressively understood as market failures within the neoliberal state. Indeed, for both 'dis-abled' and 'elderly' bodies, the "injunction to 'die sooner' is becoming hauntingly clear" (Gullette 2004, 50).

In general, supporters of the 'duty-to-die' movement argue that recent advances in biomedical technology have contributed to the 'obligation' for the elderly and the disabled to die. John Hardwig (1997, 35) explains that our current medical practices "enables many of us to survive longer than we can take care of ourselves, longer than we know what to do with ourselves, longer than we even are ourselves." In other words, the elderly are simply sitting around, waiting to die, all the while consuming precious resources that could be used for more productive—and hence, more deserving—able-bodied people. Within a society rife with ableism and ageism, dignity to the 'disabled' or 'elderly' is extended only when it comes to ending one's life. There is, in other words, dignity in death but not in life. The 'shame' of being cared for is recast as becoming a burden; something to be avoided—even if this avoidance occurs through dying (Tyner 2016, 188). Hardwig (2009, 38), for example, believes that "many of us now worry that death will come too late—long after life has lost its usefulness and its savor, long after we have ceased to have a 'life'." Consequently, following this argument, we must be 'rational' in our calculations of personal life and death. Hardwig (2009, 38) explains that today, dying well requires wisdom, skill, and prudential virtues. And thus, while one may no longer be a productive member of society, one can still do the rational thing and die. Proponents of the 'duty-to-die' movement thus attempt to downplay accusations of ageism and ableism. Supporters, for example, promote the neoliberal idea of responsible *management* of one's life. For Hardwig (2009, 39), "there is no disrespect or devaluing of the elderly in this." He explains that "it is a simple acknowledgement of the implications of being alive, and aware that we are alive" (p. 39). Much as the citizens of the world of *Logan's Run*, we must meekly—if not joyously—submit to the metaphoric hour-glasses embedded in our palms.

Tellingly, proponents of the duty-to-die movement, similar to the Deep Sleep operatives in *Logan's Run*, make wide use of the terms 'duty,' 'obligation,' and 'responsibility' (Hardwig 1997, 34). Hardwig for example proposes an ever-expansive rubric under which people are obligated to die. His recommendation extends well beyond the (legal) refusal of life-prolonging medical treatment and the completion of advance directives refusing life-prolonging treatment. Indeed, Hardwig (1997, 35) suggests that "a duty to die can go well beyond that. There can be a duty to die before one's illnesses would cause death, even if treated only with palliative measures." He concludes: "*there may be a fairly common responsibility to end one's life in the absence of any terminal illness at all*" and that "*there can be a duty to die when one would prefer to live*" (p. 35). Let us be perfectly clear as to the implications of Hardwig's position. People *not* suffering from a terminal illness, who want to live, have an obligation—a responsibility—to die. Here, simply deciding to live is re-fashioned as being 'irresponsible': one's mere existence becomes a burden to others. It is but a small step to the initiation of perverted 'life-tables' that schedule when one is called upon to 'voluntarily' be euthanized.

Unlike the narrative of *Logan's Run*, however, contemporary neoliberal and neoconservative ideologies obfuscate the potential ageist and ableist attitudes that surround 'mercy killing' and the legalization of physician-assisted suicide. Martin Gunderson (2004, 37) for example supports both physician-assisted suicide and the 'responsibility' to refuse life-prolonging treatments least one becomes a burden on others. He notes that numerous studies have shown how caring for family members with such conditions as severe depression, Alzheimer's disease, traumatic brain injury, and vascular dementia can lead to anxiety, depression, social isolation, physical illness, as well as financial problems. To this end, Gunderson concludes that these afflicted bodies should refuse to become a burden and thus refuse to live.

The specter of climate change has, ironically, provided further impetus to a neoliberal and conservative ideology to 'sacrifice' those unable to contribute *productively* to society. With the pronouncement of the dawn of the Anthropocene, scientists have argued we are inhabiting a moment of unprecedented human disruption of natural environments and habitats (Ojeda, Sasser, and Lunstrum 2020, 316). To this end, the Anthropocene presumes that human population growth, resource extraction, and resource consumption practices have made human beings the primary force behind planetary geologic and ecological change (Ojeda, Sasser, and Lunstrum 2020, 316; see also Crutzen and Stoermer 2000). Reminiscent of the threat of overpopulation that motivated Nolan and Johnson in writing *Logan's Run*, the fear underlying environmental collapse in the Anthropocene has been coopted by those seeking to re-engineering life itself on a planetary scale. Frequently, these attempts are manifest in the forwarding of coercive measures that seek to minimize population growth, such as sterilization (Hendrixson et al. 2020; Ojeda, Sasser, and Lunstrum 2020; Shaw and Wilson 2020). However, there is also a growing trend to encourage other, end-of-life practices, notably euthanasia, to stem the so-called tide of unbridled population growth.

One's obligation to die is, ultimately, predicated upon a veiled ageist and ableist market logic whereby a meaningful life is revealed through the dictates of capital. That is, proponents of duty-to-practices justify such approaches on the grounds that it is necessary to 'let die' a significant portion of the population lest society succumb to the catastrophic consequences of overpopulation. Hardwig (1997, 40), for example, concedes that the "claim that there is a duty to die will seem to some a misplaced response to social negligence." He admits that "if our society were providing for the debilitated, the chronically ill, and the elderly as it should be, there would be only very rare cases of a duty to die" (p. 40). But, Hardwig concludes, this is not the case and, hence, he is "asking the sick and debilitated to step in and accept responsibility because society is derelict in its responsibility to provide for the incapacitated" (p. 40). One is hard-pressed to fathom the logic of Hardwig's argument. Because society has failed to address social injustices—presumably because these are not profitable—those who are redundant should commit suicide. Indeed, proponents of euthanasia often fail to acknowledge the inequalities and injustices of resource extraction and consumption that mark—or mar—a planet rendered vulnerable in the Anthropocene. To this end, the elderly and the disabled are no longer perceived to embody any worth—as defined by market rationality. They are presumably beyond any productive function for our modern, post-industrial economy. The ageism implicit in this sentiment is all-the-more palpable when juxtaposed with the threat of diminishing resources. As Walter Glannon (2002, 282) explains,

> a world in which everyone's life span was extended beyond 85 years might increase competition for scarce resources and lower the quality

of life for all. The moral and prudential argument against this would be that a longer life for all would mean the defeat of people's interest in having good lives. In this regard, living longer might be not just undesirable but bad.

Conclusions

In Nolan and Johnson's fictional account, a series of crises culminated in a pervasive state apparatus that determined who was to live—and for how long. The 'duty-to-die' movement, likewise, articulates a vision of a future society whereby persons deemed redundant, that is, non-productive, are obliged to sacrifice for the collective good of society. Is *Logan's Run* beyond the realm of possibility? Or does the novel—when read against the grain of the 'duty-to-die' movement—portend a politics of becoming?

In late December 2019, a cluster of pneumonia cases, caused by a newly identified coronavirus, was documented in Wuhan, China. Patients exhibited a variety of symptoms, including fever, coughs, fatigue, shortness of breath, and sore throats. The elderly and those with underlying medical conditions, such as hypertension, pulmonary disease, diabetes, and cardiovascular disease, appeared to be especially at risk (Guo et al. 2020). In the United States, the first case of COVID-19 was confirmed on January 21, 2020 and ten days later, the first case of human-to-human coronavirus transmission in the country was confirmed (Schumaker 2020). Within months, the virus would killed over a hundred thousand people in the United States alone. In short order, health-care services and workers became overstretched and mortuaries overflowed; in the worst cases—both in the United States and elsewhere—the dying and the dead were abandoned in their homes, in care homes, or on the streets (Maddrell 2020).

On January 30, 2020 WHO declared a 'public health emergency of international concern,' a designation reserved for extraordinary events that threaten to spread internationally (Schumaker 2020). However, as the virus began its steady, inexorable sweep across the globe—and in the process infecting not only people but the global economy—conservative politicians in the United States sat by idly. Scientific experts, long chastised by President Trump and his followers as members of the 'Deep State,' were silenced and subdued. Fearing his re-election chances, President Trump downplayed the pandemic. On January 22, he received his first question about the coronavirus during an interview on CNBC. Asked whether he was worried about a potential pandemic, Trump dismissed the question, replying "No. Not at all. And we have it totally under control. It's one person coming in from China.... It's going to be just fine" (quoted in Abutaleb et al. 2020). Thus, instead of marshaling resources in anticipation of critical shortages of medical supplies, ventilators, and PPE, he dismissed warnings issued by the Centers for Disease Control (CDC) and the WHO with false equivalences, suggesting, for example, that COVID-19 was comparable to the seasonal flu.

To this end, Trump politicized the pandemic, claiming that the disease was a 'hoax' perpetrated by his 'enemies.' Notably, in December 2019 Trump was impeached by the Democratic-controlled House of Representatives; and although the Republican-controlled Senate acquitted Trump in the first week of January, Trump framed COVID-19—when he acknowledged it at all—as simply another weapon in the arsenal of his political opponents. This would prove a dangerous narrative in the coming months.

By late February Trump could no longer simply ignore COVID-19. His response, however, remained politically charged. On February 26, 2020—as the CDC confirmed the 60th case in the United States—President Trump designated his Vice-President, Mike Pence, to lead the U.S. response to the coronavirus (Schumaker, Winsor, and Pereira 2020). During the press conference, President Trump continued to send mix messages, however, repeatedly saying that the risk to Americans was low. Noting that tens of thousands of people in the United States die every year from influenza, Trump minimized the risk of COVID-19. "You have 15 people," the president explained, "and the 15 within a couple of days is going to be down to close to zero" (quoted in Chait 2020). President Trump dithered, "You don't want to see panic, because there's no reason to be panicked about it. There's a chance it won't spread, too. And there's a chance it will" (quoted in Cathey and Gittleson 2020). Pence echoed the remarks of the president, stating simply, "The threat to the American public remains low" (quoted in Schumaker, Winsor, and Pereira 2020).

Public health experts at the CDC evinced considerably more concern than either President Trump or Vice-President Pence. As the top officials in the U.S. government minimized the threat of COVID-19 to the American public, health officials watched with growing consternation the deadly effects of the disease in Italy, Iran, and South Korea. Dr. Nancy Messonnier, director of the U.S. Centers for Disease Control and Prevention's National Center for Immunization and Respiratory Diseases, explained in a news conference that "Ultimately, we expect we will see community spread in this country" (quoted in Winsor, Schumaker, and Nathanson 2020). The message was clear. Infections, at the moment, remained low; but this could provide the necessary lead time to prepare for the eventual spread of the disease. Trump, however, fumed at what he considered impertinence by public health experts. Consequently, in response to the comments made by Dr. Messonnier, Trump marginalized the CDC, effectively muzzling the dissemination of any information that went against his claims that the virus was under control (Roberts 2020).

As the weeks turned into months, Trump's stubborn unwillingness to adequately respond to the growing threat hardened and the number of infections and deaths mounted. Medical experts cautioned that immediate and drastic responses were necessary: mass production of medical supplies and the establishment of an efficient distribution system; the initiation of protocols for widespread testing and contact tracing; and the implementation of 'shelter-in-place' orders to minimize community transmission.

Trump however refused to initiate comprehensive plans at the federal level, preferring instead to 'delegate' responsibility to the state level in a thinly veiled attempt to deflect any potential blame from his mishandling of the crisis. In addition, the system *as a whole* revealed itself to be unable to provide adequately in the face of crisis. Simon Springer (2020, 114) is blunt in his judgment: "Profit over people is the true pandemic."

The numbers indeed were staggering. On May 27, the official death toll in the United States surpassed 100,000 (Lim 2020). However, the public claims of President Trump that the pandemic was unprecedented and that he had no warning were easily disproven. In fact, as subsequent investigative reports documented, the warning signals were apparent for a long time. The Trump administration received its first formal notification of the coronavirus outbreak in China on January 3 and within days, U.S. intelligence agencies were signaling the seriousness of the threat to Trump by including a warning about the coronavirus in his daily briefings (Abutaleb et al. 2020). Trump though prevaricated for over two months, squandering precious time as the death toll continued to rise.

State and local officials stepped into the political void, albeit in a haphazard and geographically uneven manner. Several governors and mayors—usually, but not always Democrats—issued mandatory shelter-in-place regulations and issued other guidelines, including social distancing and the wearing of masks. However, while the limited shelter-in-place regulations, enacted sporadically across the country, proved insufficient to stem the COVID-19 tide, these measures did exact a devastating toll on the economy. Within weeks, millions of Americans found themselves unemployed and without health care; Congress, in response, were negligible in their ability to respond to those people hurt most by the deepening economic crisis.

In *Logan's Run*, the apocalypse unfolded via a series of interconnected crises. In similar fashion, in the United States the failed government response to the pandemic ushered a devastating economic crisis which, in turn, fueled existent political divisions and revealed ever more clearly the fractures and fault lines of society. In April, for example, several 'Astro-Turf' social movements erupted, mostly in key swing states or states with Democratic governors, demanding an end to shelter-in-place rules and a return to 'normal.' Protesters, mostly white, frequently armed, and often waving Confederate flags, decried what they believed was an abrogation of their personal freedoms (Jeffery 2020). Trump, for his part, joined the chorus of white protestors, simultaneously downplaying the serious of the pandemic and elevating fears of vanishing liberties (Bianco 2020; Valasik 2020). Outside the state capital in Olympia, Washington, for example, thousands of protestors gathered to call for the end of stay-at-home orders. Protestors carried American flags and Trump flags, gun rights advocates brought their firearms, and several held posters that read 'Give Me Liberty or Give Me COVID-19' and Republican Representative Robert Sutherland stoked the crowd with demands to 'take back the state' (Sorace 2020).

When health experts warned that a premature relaxation of precautionary measures would result in the premature death of thousands of people, protestors—along with Trump and his spokespersons—evinced a utilitarian ethos that the 'cure' cannot be worse than the 'illness.' As Dr. Anthony Fauci, chief epidemiologist of the National Institute for Health, remarked, "How many deaths and how much suffering are you will to accept to get back to what you want to be, some form of normality, sooner rather than later?" (quoted in Smith 2020). Little empathy however was expressed by Republic politicians or by the Trump-supporting protestors toward those who were dying—especially the elderly, but also minorities and those with disabilities and comorbidities (see for example Carrieri, Peccatori, and Boniolo 2020; Laster Pirtle 2020). Across the United States Republican officials relaxed or ended public safety protections in their states; many politicians, for example, abandoned or declined to extend shelter-in-place orders and refused to enact mandatory guidelines to wear masks and to socially distance in public places (Smith 2020).

Seemingly incredulously, many conservative politicians have been upfront in their callous disregard for life and their willingness to let die the sick and the elderly. In late April, for example, Lt. Governor Dan Patrick (R-TX) rationalized that there are more important things than living, and that it was incumbent that "we've got to take some risks and get back in the game and get this country back up and running" (quoted in Benen 2020). More precisely, Patrick called upon senior citizens to 'sacrifice' their lives in order to maintain the American way of life for future generations. He explained,

> If that's the exchange, I'm all in. That doesn't make me noble or brave or anything like that. I just think that there are a lot of grandparents … that we most care about are those children. I want to live smart and see through this, but I don't want to see the whole country sacrificed.
>
> (quoted in Pierce 2020)

Chris Christie, former governor of New Jersey, flippantly remarked, "Of course, everybody wants to save every life they can—but the question is, towards what end, ultimately?" (quoted in Smith 2020). And an editorial in *The Federalist* was even more blunt:

> Of course, it sounds very callous to talk about considering the costs. It seems harsh to ask whether the nation might be better off letting a few hundred thousand people die. Probably for that reason, few have been willing to do so publicly thus far. Yet honestly facing reality is not callous, and refusing even to consider whether the present response constitutes an even greater evil than the one it intends to mitigate would be cowardly…. The current response is quickly driving the United States into a recession, which will result in a great deal of misery for tens of

> millions of people. Again, balancing lives against money sounds harsh, but everyone does so—and must do so—whether he is conscious of the fact or not.
>
> (quoted in Pierce 2020)

The message was clear: the demand to patronize bars and bowling alleys, barber shops, and beauty parlors outstripped the need to save the sick and the elderly. As Dana Milbank (2020) writes, the Trump administration essentially offered senior citizens as sacrifices on the altar of economic recovery. Indeed, for Trump and his Republic acolytes, the tradeoff in reopening the economy was *always* going to be paid by the most vulnerable. According to their necro-political reasoning—itself a toxic mixture of neo-Malthusianism and social Darwinism—those who succumb to the disease—notably the elderly or those with 'comorbid' conditions—would die anyway. All that remains is the establishment of Sleepshops for the superfluous to report for duty.

Ultimately, the response of the Trump administration portends a dangerous new world where environmental catastrophes—wrought by a rapacious capitalism—are likely met with brutal population control measures, including the sacrifice of the elderly or those broadly conceived as unproductive. In the process, inequalities in resource scarcities and consumption are normalized along the dictates of unfettered capital accumulation and the suffering of the impoverished twisted in justification of coercive practices to curb the neo-Malthusian ghost of overpopulation. To this end, we readily see the embrace of violent inaction among government officials who are willing to 'let die' those whom they determine are unproductive and thus disposable.

Near the end of his life, former Vice President Hubert H. Humphrey said that "the ultimate moral test of any government" is how it treats the most vulnerable, notably "those in the twilight of life, our elderly" (quoted in Milbank 2020). In the United States, seniors are already largely an afterthought (Jones 2020). Indeed, as Jones (2020) documents, senior security is higher in the United States than in many other wealthy countries: half of all seniors who live alone lack the means to cover basic expenses and many low-income seniors continue to work, or enter poorly regulated care facilities that pose unique dangers to their health. Jones (2020) concludes:

> As coronavirus carves through the elderly, it tells us something ugly about the high price of the American project. The prosperity it engenders is real but limited; it is exclusionary by design. Wealth flows upward, where it stays, and creates an inverted pyramid that bloats at the top then vanishes to a fine point at the bottom. Proper care for the elderly and for people with disabilities requires what some corporate executives might call a restructuring—an unpalatable task for those already at the top.

In light of Humphrey's acumen, we readily see how apocalyptic politics are revelatory, exposing deeply held beliefs of what is considered meaningful. Vitally, these discernments provide insight into the most fundamental of bio- and necro-political questions; precisely, who lives, who dies, and who decides. Arguably, within our present moment, the collective answer is simply, 'who profits?'

Notes

1 In 1976 a film version based on a screenplay written by David Zelag Goodman was released. This chapter is based exclusively on Nolan and Johnson's novel.
2 Similar dystopian futures of ageist euthanasia are depicted in Anthony Trollope's (1882) novel *The Fixed Period* and Margaret Atwood's (2014) short story "Torching the Dusties."
3 In the 1976 film version, the age was set at 30.

References

Abutaleb, Y., Dawsey, J., Nakashima, E., and Miller, G. (2020) The U.S. Was Beset by Denial and Dysfunction as the Coronavirus Raged. *The Washington Post*, April 4, 2020 at www.washingtonpost.com/national-security/2020/04/04/coronavirus-government-dysfunction/?arc404=true (accessed June 27, 2020).

Ahn, S. (2020) Old Age and Euthanasia in Anthony Trollope's The Fixed Period. *Neophilologus* 104(3):439–451.

Atwood, M. (2014) Torching the Dusties. In Atwood, M. (ed.), *Stone Mattress: Nine Tales*. London: Bloomsbury, 225–270.

Bauman, Z. (2004) *Wasted Lives: Modernity and Its Outcasts*. Malden: Polity.

Benen, S. (2020) Texas Dan Patrick: There Are More Important Things Than Living. *MSNBC*, April 22, 2020 at www.msnbc.com/rachel-maddow-show/texas-dan-patrick-there-are-more-important-things-living-n1189811 (accessed June 8, 2020).

Bettini, G. (2013) Climate Barbarians at the Gate? A Critique of Apocalyptic Narratives on 'Climate Refugees'. *Geoforum* 45: 63–72.

Bianco, M. (2020). Michigan Coronavirus Protestors Shout 'Liberty!'—as Right-Wing Rhetoric Weaponizes Freedom. *NBC News*, May 14, 2020 at www.nbcnews.com/think/opinion/michigan-coronavirus-protesters-shout-liberty-right-wing-rhetoric-weaponizes-freedom-ncna1206596 (accessed June 8, 2020).

Cacho, L. M. (2012) *Social Death: Racialized Rightlessness and the Criminalization of the Unprotected*. New York: New York University Press.

Carrieri, D., Peccatori, F. A., and Boniolo, G. (2020) COVID-19: A Plea to Protect the Older Population. *International Journal for Equity in Health* 19: 72.

Cathey, L. and Gittleson, B. (2020) Pence to Lead US Response to Coronavirus as Trump Urges Americans to Prepare. *ABC News*, February 26, 2020 at https://abcnews.go.com/Politics/trump-announces-news-conference-coronavirus-cdc-days-mixed/story?id=69228747 (accessed June 27, 2020).

Chait, J. (2020) Trump: I Was Right, Coronavirus Cases 'Will Go Down to Zero, Ultimately'. *New York Magazine*, April 28, 2020 at https://nymag.com/intelligencer/2020/04/trump-coronavirus-cases-will-go-down-to-zero-ultimately.html (accessed June 27, 2020).

Cholbi, M. (2010) The Duty to Die and the Burdensomeness of Living. *Bioethics* 24: 412–420.
Commons, J. R. (1907) *Races and Immigrants in America*. New York: Macmillan.
Corlett, J. A. (2001) Is There a Moral Duty to Die? *Health Care Analysis* 9: 41–63.
Crutzen, P. and Stoermer, E. (2000) The 'Anthropocene'. *Global Change Newsletter* 41: 17–18.
Davis, T., Isakjee, A., and Dhesi, S. (2017) Violent Inaction: The Necropolitical Experience of Refugees in Europe. *Antipode* 49(5): 1263–1284.
Dittmer, J. and Spears, Z. (2009) Apocalypse, Now? The Geopolitics of Left Behind. *GeoJournal* 74: 183–189.
Dore, M. K. (2011) Physician-Assisted Suicide: A Recipe for Elder Abuse and the Illusion of Personal Choice. *The Vermont Bar Journal* (Winter): 1–4.
Ekland-Olson, S. (2012) *Who Lives, Who Dies, Who Decides: Abortion, Neonatal Care, Assisted Dying, and Capital Punishment*. New York: Routledge.
Emmerson, P. (2019) From Coping to Carrying On: A Pragmatic Laughter between Life and Death. *Transactions of the Institute of British Geographers* 44(1): 141–154.
Enck, R. E. (2010) Physician-Assisted Dying. *American Journal of Hospice & Palliative Medicine* 27: 441–443.
Fenn, R. (2003) Apocalypse & the End of Time. *Dædalus* 132(2): 108–112.
George, R.J. D., Finlay, I. G., and Jeffrey, D. (2005) Legalized Euthanasia Will Violate the Rights of Vulnerable Patients. *British Medical Journal* 331: 684–685.
Glannon, W. (2002) Identity, Prudential Concern, and Extended Lives. *Bioethics* 16(3): 266–283.
Golden, M. and Zoanni, T. (2010) Killing Us Softly: The Dangers of Legalized Assisted Suicide. *Disability and Health Journal* 3: 16–30.
Gullette, M. M. (2004) *Agewise: Fighting the New Ageism in America*. Chicago: University of Chicago Press.
Gunderson, M. (2004) Being a Burden: Reflections on Refusing Medical Care. *Hastings Center Report* 34: 37–43.
Guo, Y. R., Cao, Q. D., Hong, Z. S., Tan, Y. Y., Chen, S. D., Jin, H.J.,… & Yan, Y. (2020). The Origin, Transmission and Clinical Therapies on Coronavirus Disease 2019 (COVID-19) Outbreak–an Update on the Status. *Military Medical Research* 7(1): 1–10.
Hardwig, J. (1997) Is There a Duty to Die? *Hastings Center Report* 27: 34–42.
Hardwig, J. (2009) Going to Meet Death: The Art of Dying in the Early Part of the Twenty-First Century. *Hastings Center Report* 39: 37–45.
Hendrixson, A., Ojeda, D., Sasser, J. S., Nadimpally, S., Foley, E. E., & Bhatia, R. (2020). Confronting populationism: Feminist challenges to population control in an era of climate change. *Gender, Place & Culture*, *27*(3): 307–315.
Jeffery, A. (2020) Scenes of Protests across the Country Demanding States Reopen the Economy amid Coronavirus Pandemic. *CNBC*, April 18, 2020 at www.cnbc.com/2020/04/18/coronavirus-scenes-of-protests-across-the-country-demanding-states-reopen-the-economy.html (accessed June 27, 2020).
Jones, S. (2020) No One Should be Surprised that America Abandoned the Elderly to Die. *New York Magazine*, July 9, 2020 at www.nymag.com/intelligencer/2020/07/america-is-sacrificing-the-elderly-to-coronavirus.html (accessed July 28, 2020).
Laster, P. and Whitney N. (2020) Racial Capitalism: A Fundamental Cause of Novel Coronavirus (COVID-19) Pandemic Inequities in the United States. *Health Education & Behavior*. DOI: 10.1177/1090198120922942

Laurie, E. W. and Shaw, I. G. R. (2018) Violent Conditions: The Injustices of Being. *Political Geography* 65: 8–16.

Lawson, V. (2007) Geographies of Care and Responsibility. *Annals of the Association of American Geographers* 97(1): 1–11.

Lim, C. (2020) More Than 100,000 People Have Died of the Coronavirus in the US. *BuzzFeed.News*, May 27, 2020 at www.buzzfeednews.com/article/clarissajanlim/coronavirus-death-toll-100000-us (accessed June 27, 2020).

Maddrell, A. (2020) Bereavement, Grief, and Consolation: Emotional-Affective Geographies of Loss during COVID-19. *Dialogues in Human Geography* 10(2): 107–111.

McIntyre, M. and Nast, H. J. (2011) Bio(Necro)Polis: Marx, Surplus Populations, and the Spatial Dialectics of Reproduction and 'Race'. *Antipode* 43(5): 1465–1488.

McMahan, J. (2002) *The Ethics of Killing: Problems at the Margins of Life*. New York: Oxford University Press.

Milbank, D. (2020) America's Seniors, Sacrificed on the Altar of Reopening. *The Washington Post*, May 22, 2020 at www.washingtonpost.com/opinions/2020/05/22/americas-seniors-sacrificed-altar-reopening (accessed July 28, 2020).

Mussgnug, F (2012) Naturalizing Apocalypse: Last Men and Other Animals. *Comparative Critical Studies* 9(3): 333–347.

Nolan, W. F. and Johnson, G. C. (1967) *Logan's Run*. New York: Bantam Book.

Ojeda, D., Sasser, J. S., and Lunstrum, E. (2020). Malthus's Specter and the Anthropocene. *Gender, Place & Culture* 27(3): 316–332.

Pierce, C. P. (2020) No Thanks, I Will Not Immolate Myself on the Altar of Your Stock Portfolio. *Esquire*, March 24, 2020 at www.esquire.com/news-politics/politics/a31910755/coronavirus-republicans-sacrifice-old-people-stocks (accessed July 28, 2020).

Pitetti, C. (2017) Uses of the End of the World: Apocalypse and Postapocalypse as Narrative Modes. *Science Fiction Studies* 44(3): 437–454.

Roberts, W. (2020) Trump Sidelines CDC in Push to Reopen the US Economy. *Aljazeera*. May 20, 2020 at www.aljazeera.com/news/2020/05/trump-sidelines-cdc-push-reopen-economy-200520191135513.html (accessed June 8, 2020).

Schlosser, K. (2015) Apocalyptic Imaginaries, Gramsci, and the Last Man on Earth. *GeoHumanities* 1(2): 307–320.

Schumaker, E (2020) Timeline: How Coronavirus Got Started. *ABC News*, July 28, 2020 at https://abcnews.go.com/Health/timeline-coronavirus-started/story?id=69435165 (accessed August 30, 2020).

Schumaker, E., Winsor, M., and Pereira, I. (2020) Latest American Infected with Coronavirus Has No Relevant Travel History: CDC. *ABC News*, February 26, 2020 at https://abcnews.go.com/International/us-military-coronavirus-patient-cases-surge-italy-south/story?id=69225004 (accessed June 27, 2020).

Shaw, A. and Wilson, K. (2020) The Bill and Melinda Gates Foundation and the Necro-Populationism of 'Climate-Smart' Agriculture. *Gender, Place & Culture* 27(3): 370–393.

Shaw, I. G. R. and Waterstone, M. (2019) *Wageless Life: A Manifesto for a Future Beyond Capitalism*. Minneapolis: University of Minnesota Press.

Smith, J. (2020) They Are Giving You Death and Calling it Liberty. *Rolling Stone*, May 6, 2020, at www.rollingstone.com/politics/political-commentary/coronavirus-reopen-states-death-liberty-trump-994421/ (accessed July 28, 2020).

Snaith, H. (2017) Dystopia, Gerontology and the Writing of Margaret Atwood. *Feminist Review* 116: 118–132.

Sorace, S. (2020) Protestors Rally against Washington's Coronavirus Stay-at-Home Order: 'Give Me Liberty or Give Me Covid 19'. *Fox New*, April 20, 2020 at www.foxnews.com/us/washington-coronavirus-protest-inslee-stay-at-home (accessed June 27, 2020).

Springer, S. (2020) Caring Geographies: The COVID-19 Interregnum and a Return to Mutual Aid. *Dialogues in Human Geography* 10(2): 112–115.

Swyngedouw, E. (2012) Apocalypse Forever? Post-Political Populism and the Spectre of Climate Change. *Theory, Culture & Society* 27(2–3): 213–232.

Till, K. E. (2012) Wounded Cities: Memory-Work and a Placed-Based Ethics of Care. *Political Geography* 31(1): 3–14.

Trollope, A. (1882) *The Fixed Period.* Oxford: Oxford University Press.

Tyner, J. A. (2012) *Genocide and the Geographical Imagination: Life and Death in Germany, China, and Cambodia.* Lanham: Rowman & Littlefield Publishers.

Tyner, J. A. (2016) *Violence in Capitalism: Devaluing Life in an Age of Responsibility.* Lincoln: University of Nebraska Press.

Tyner, J. A. (2019) *Dead Labor: Toward a Political Economy of Premature Death.* Minneapolis: University of Minnesota Press.

Valasik, M. (2020) Why Are White Supremacists Protesting to 'Reopen' the U.S. Economy. *Salon,* April 30, 2020 at www.salon.com/2020/04/30/why-are-white-supremacists-protesting-to-reopen-the-us-economy_partner/ (accessed June 27, 2020).

Winsor, M., Schumaker, E., and Nathanson, M. (2020) CDC Warns Americans of 'Significant Disruption' from Coronavirus. *ABC News*, February 25, 2020 at https://abcnews.go.com/International/coronavirus-live-updates-cdc-warns-americans-significant-disruption/story?id=69190968 (accessed June 27, 2020).

8 The catastrophic drive

Lucas Pohl and Samo Tomšič

The narration of the apocalypse is probably more widespread than any other kind of storytelling in our time. It unfolds in various forms and influences the temperament of our era: from Hollywood blockbusters and science fiction novels, TV documentaries and video games to blog posts, art projects, journalistic reports and edited volumes like this one. In the first weeks after COVID-19 began to dominate the public media, political debates and everyday life in many parts of the world, it almost seemed for a moment as if the end of the world had finally come. The images of overflowing hospitals in Italy, lockdowns of whole suburbs in Spain, queues in front of gun shops in the USA and empty shelves in grocery stores all over the world – these images which most of us are familiar with from various kinds of popular culture suddenly started to spread all over the daily press and social media. All of a sudden, the fictive scenarios of all the stories about the end times, published and spread by the entertainment industry over the last decades, seemed to finally become reality. The "coronapocalypse" began trending in social media and it certainly captured the spirit of this moment, namely, a situation that is so utterly out of joint that it is hard to imagine how things could go back to normal. Although these developments did have an unexpected side effect, since, at least for a moment, they seemed to have questioned the well-known slogan attributed to Frederic Jameson (2003: 76), "it is easier to imagine the end of the world than the end of capitalism". The first weeks of the pandemic may at first have demonstrated the validity of this statement, namely how surprisingly easy it is to imagine the end of the world (compared to the end of capitalism).[1] But the registered environmental effects of the temporary lockdown and the economic measures adopted at least by some European states have also shown that the neoliberal dogma "there is no alternative" can much more easily be overcome than the established capitalist imaginary permits.

A key motif underlying the apocalyptic imagination is that the end of the world not only means a physical destruction of the planet, but also, and perhaps more prominently, means an end of "the social" and thus of morality,

DOI: 10.4324/9781003128854-9

justice and reason. If there is one lesson to be learned from all the zombie apocalypse stories that have been told over the last years, it is that the real "enemy" in these stories is not the zombie but our fellow human being. Of course, the zombie is an obvious threat, an external danger, but what is even more threatening is the secret danger, the internal threat that emanates from the fellow humans. The basic premise underlying this narration is that the end of the world leads to an end of the state and the legal system, and therefore it confronts us with a return to a kind of Hobbesian "state of nature", where humans no longer live bound to the restrictions and limitations of society.[2] The apocalypse presents a world in which there are no laws except for the "laws of nature", a world dominated by a "war of all against all". In the first weeks of the COVID-19 pandemic, this vision of an end of society acquired a strange actuality. In fact one of the reasons why it was so easy to imagine the end of the world at that time was that one could actually witness a certain crumbling of the social. Whether it's consumers fighting over toilet paper, countries refusing medical aid to each other, or extreme populists such as Bolsonaro or Trump pushing for a return to business as usual, all these examples suggest that the war of all against all does not lie in some distant fantasmatic past, but virtually exists here and now, intensifying its reality in the crisis-driven situation. The recent developments are no exception; they confront us with an end of the social and the rise of the "asocial". Zygmunt Bauman (2007: 88) captures this proto-Hobbesian notion of the apocalypse in a nutshell by stating that society is necessary because otherwise humans would end up carrying only about their personal well-being, "which would guide individual conduct towards the wasteland of asociality if it were not constrained, trimmed and counterbalanced by the power-assisted and authority-operated 'reality principle'".[3] The basic assumption here is that society, as the sphere of the social, is necessary because it prevents us from turning into asocial beings. When the realm of society begins to lose its grip, as it was happening throughout the first weeks of the pandemic, it therefore leads to the negation of the social and paves the way to the "wasteland of asociality".

But does Bauman's line not grasp very precisely the neoliberal doctrine, as it was formulated by Margaret Thatcher's notorious statement that "There is no such thing as society"? The asocial is already here, in the midst of the social, in the very economic doctrines that are violently implemented and put into practice for the price of the progressive dissolution of the social. Further, from this viewpoint, the asocial does not describe individual characters and actions but rather the structural tendencies and developments of the capitalist socioeconomic order itself. In this chapter, we aim to take a closer look at the notion of asociality underlying capitalism. In contrast to the way of imagining the end of the world outlined above, this will allow us to consider the asocial not as the counterpart nor "outside" of society, but rather as playing a crucial role "for" the social itself. It is only through the asocial that we can properly understand how society, and with society we mean *capitalist* society, works. While it is not difficult to imagine the apocalypse as some sort of "big bang" catastrophe that puts an end to the social, it is more difficult to take the catastrophic dimensions into account that are immanently at work within our capitalist everyday lives. Capitalism pursues production for its own sake, as Marx argues, self-sufficient production, which serves no other purpose than the augmentation of profit. Behind the appearance of usefulness, there is the imperative of uselessness at work within capitalism, and behind the appearance of the social, there lies the dimension of the asocial. Considered as "overproduction", a production for its own sake, capitalism can be at best described as an asocial mode of production (rather than a social one), we argue. Capitalism is a system, which always and inevitably comprises the negation of the social and thus incorporates what Freud called the "death-drive". In the following, we draw the consequences from this argument by formulating an approach that aims to understand the end of the world as a gradual "surplus product" of capitalism, as the flipside of growth-oriented economy and the self-valorisation of capital.

Freudian-Marxist lessons

Freud first introduced the notion of death drive in *Beyond the Pleasure Principle* (1920). The notion made many psychoanalysts raise their eyebrows and moreover deeply split the psychoanalytic community. But the "death-drive hypothesis" was meant to fulfil a double function. On the one hand, it should explain individual psychic phenomena such as compulsive repetition of unpleasurable and traumatic events, aggressiveness and neurotic guilt, but also the peculiar intertwining of pleasure and unpleasure, such as notably in sadism and masochism. On the other hand, the death drive quickly turned out to be a handy tool for interpreting the destructive tendencies that underpin the capitalist cultural condition, as it was later elaborated by Freud in essays like *Civilization and Its Discontents* (1929). With this shift from individual to social, Freud proposed a radical and rather disturbing view of culture, which points back to one specific mode of the drive's satisfaction, *die Wendung gegen die eigene Person*, the act of turning against oneself (Freud 2000: 90).[4]

It seems quite symptomatic that Freud's theory of drives took this dramatic turn in a most critical moment of history. The work on *Beyond the Pleasure Principle* began in March 1919, in the immediate aftermath of the First World War. With this unprecedented type of warfare and other overtly destructive features of modern life, a new pathology began to proliferate, traumatic neurosis. It is also significant that Freud's text was written during the greatest pandemic of the twentieth century. Caused by an influenza virus, which began spreading across the globe in spring 1918, the pandemic amounted to an estimated death of 50 million worldwide, including Freud's daughter Sophie.[5] This historical and biographical contextualisation is not meant to downplay the weight of this groundbreaking shift in Freud's work. Despite the nexus of personal and political, which may have partly motivated Freud's confrontation and working through the traumatic impact of war, crisis and death, the more interesting part of Freud's diagnostic lies elsewhere. The hypothesis of the death drive is elaborated in an epoch, whose overall atmosphere bears uncanny resemblance to our catastrophic times. Or perhaps better, it anticipates our crisis-driven present, in which destructive developments such as climate breakdown, economic instability, the return of racism and sexism at the centre of political life, global warfare and most recently the COVID-19 pandemic expose *the* problem that occupied Freud's later work, the intimate link between culture and traumatism, or more specifically, capitalism as organised, systemic traumatism, a system running amok.[6]

Although Freud's cultural diagnostic moves in an entirely different framework, its structural aspects intersect with a suspiciously similar movement in Marx. From the psychoanalytic viewpoint, it is significant that the term *Trieb* surfaces, whenever Marx makes an important step in his discussion of the dynamic of capitalist abstractions (money, value and capital) or of the contradiction between capital and labour. The drive enters the stage in the guise of drive for enrichment (*Bereicherungstrieb*), drive of accumulation (*Akkumulationstrieb*), drive of self-valorisation (*Selbstverwertungstrieb*) and even the "drive for an unlimited extension of the working day".[7] With this move from enrichment, which can still be associated with individuals, to self-valorisation, which directly concerns the immanent dynamic of value, Marx draws attention to the underlying structural contradiction between society and subjectivity, on the one hand, and the tendencies of capitalist economy, on the other.

Unlike his counterparts in political economy, who at first glance identified similar passions, motivations and driving forces in economic subjects – most notably Adam Smith with his notion of private interest – Marx in advance rejects the simplistic view, according to which the drive is a psychological feature of individuals. It may appear as such in the figure of the miser and the capitalist alike. But this personalised façade should not seduce us into overlooking that, in difference to the miser, the capitalist fulfils the task of socially administrating the structural tendencies of capital. Marx's move from the individuating phenomenology of greed to the structural tendencies of the capitalist mode of production encounters the same compulsion as Freud, and he explains this compulsion by means of an impersonal force pertaining to economic abstractions and processes themselves. In doing so, Marx thus undertook a thorough depsychologisation of the drive. Understood as structural rather than psychological force, the drive thoroughly questions the assumed freedom and autonomy of economic subjects and moreover accounts for a compulsive action that dominates individual lives, minds and bodies, but also intersubjective, social relations.

On the background of this conceptual decision, Marx broke with classical political economy. As already mentioned, Adam Smith formulated a hypothesis, which only appears to point in the same direction and which obtained an ideological boost in the decades of neoliberalism, the hypothesis of human being's natural inclination to egoism. For an inattentive eye, Smith's notion of private interest could appear as some kind of conceptual predecessor of the drive. In fact, it is its mystification. In the apologetic of capitalism greed played an ambiguous role: on the one hand, it was considered a natural force reflecting human striving for self-preservation, and on the other hand, it appeared as an eminently cultural or social force, which was presumably capable of sustaining stable social relations. Greed is a central component of capitalist naturalism, the point where the advocates of capitalism strive to establish the link between capitalism and human nature.[8] In this respect, capitalism always understood itself as the social order, which is most in accordance with the presumably natural inclinations of human beings.

This psychologisation of greed re-emerges in psychoanalytic and economic contexts that are otherwise critical of capitalism, such as in Otto Fenichel or John Maynard Keynes, who both undertook a critique of the capitalist drive of enrichment and its naturalisation. But they nevertheless overlooked that Freud and Marx use the concept of the drive in order to move from psychological to structural reality, away from personal to impersonal. Moreover, Marx's historical and formal analyses already show how the drive of enrichment, which has been criticised ever since Plato and Aristotle, undergoes a wide-reaching metamorphosis, once the invention of surplus value grounded the modern socioeconomic order.

Jacques Lacan took up the resonances between psychoanalysis and the critique of political economy by insisting that there is a fundamental homology between Marx's account of the function of surplus value in the drive of capital and Freud's explanation of the drive's constant striving for more enjoyment (*Lustgewinn*). In order to pinpoint this homology, Lacan coined the term "surplus enjoyment" (*plus-de-jouir*). Both surpluses, in the fields of value and enjoyment, address the objectification of the "more", which makes the drive's satisfaction a virtually endless task. Lacan's explicit thesis is that surplus value must be regarded as systemic enjoyment, and correspondingly, the capitalist system as a machine of enjoyment – with catastrophic consequences.

Enjoyment between the social and the asocial

At this point, it may be worth recalling Lacan's occasional "definition" of enjoyment, which condenses the various complications at stake on the level of the drive and its satisfaction: "Enjoyment is what serves for nothing", *ne sert à rien* (Lacan 1999: 3, translation modified). To put it in other words, enjoyment contributes nothing to the satisfaction of needs and has no usefulness or purpose except itself. Understood in this way, enjoyment means "pleasure for the sake of pleasure" – and the term "surplus enjoyment" that Lacan coined in reference to surplus value is meant to elevate the detachment of enjoyment from usefulness to the level of the concept. In the same manner, surplus value points to an essential feature of capitalism, the organisation of production around the imperative of perpetual growth and augmentation of value, in other words, around "production for the sake of production" (Marx 1990: 742). We will return to this crucial characterisation further below.

Surplus enjoyment thus stands for enjoyment characterised by uselessness. "Jouissance is waste", as Alenka Zupančič (2006: 157) puts it. This feature concerns specifically the capitalist mode of enjoyment.[9] So what does this say about the capitalist social link? Freud already drew attention to the libidinal character of social relations, in other words, that these relations must be considered as libidinal bonds.[10] If viewed through the lens of the theory of the drives, the social immediately turns out to be internally torn between consistency and dissolution. The pair of Eros and the death drive brings this tension to the point. In the guise of Eros, enjoyment supposedly forms a link. In this respect, it actively contributes to the constitution and reproduction of social links. This "socialisation" of enjoyment allows Freud to assert that Eros is "the preserver of life" (Freud 1955: 54) and "holds together everything in the world" (Freud 1955: 92) by enforcing the externalisation of narcissistic, self-related libido. But it would be wrong to see in Eros an agency of equilibrium, since Freud introduced the death drive precisely as the libidinal force, which tends towards the renewal of the ultimate equilibrium that is the inanimate state. Eros thus appears as a force striving for the preservation of disequilibrium that is the state of life. It may hold together "everything in the world", but for the price of sustaining imbalance. From the perspective of the death drive, this unstable libidinal bond tends towards its own dissolution.

Freud believed that, due to this inner contradiction of the socio-libidinal, it makes sense to keep the dualism of drives. Lacan's return to Freud famously displaced the accent on the immanent split of the drive, according to which Eros and the death drive are two faces of one and the same force, suggesting that the social link, too, must be thought in its immanent tendency to dissolution. The death drive then stands for a specific complication within Eros, the backfiring of preservation of life or rather the preservation of the life of drive regardless of the life of its individual barriers. Furthermore, the death drive names the anti-social force of social links, an immanent paradox of the social.

Lacan's homology of surplus enjoyment and surplus value suggests that surplus value – understood as enjoyment of the capitalist system[11] – is useless as well. And just like surplus enjoyment, surplus value, too, is an asocial product of social production. Marx's discussion of the shift in the status of accumulation between premodernity (the figure of the miser) and modernity (the figure of the capitalist) addresses this complication (see Marx 1990: 254–255). The main achievement of capitalism consisted in the externalisation of the drive of accumulation, which has been restricted to individual greed or mania in pre-capitalist modes of production. In this restriction the drive has been a finite force, dependent on the "embodiment" in the miser and his treasure. Capital, in turn, overcomes the boundaries of treasure and stands for the liberation of the creative potential of value in all its abstraction (Marx occasionally speaks of money's *Zeugungskraft*, procreative force, which is precisely boundless). Moreover, the success of capitalism consisted in founding an entire social mode of production on the link between destruction and growth, two aspects of one and the same uselessness: surplus value can only be extracted from natural resources under the condition of environmental destruction; and it can only be extracted from the living bodies under the condition of their consumption and exhaustion, so again destruction. While it seems as if capitalism embedded in the social is something that still appeared asocial in the case of the miser, it actually liberated the asocial, amounting to a social and economic order, in which the social is an extension of the asocial (just like for Freud the reality principle is an extension of the pleasure principle).

Sustaining the self-valorisation drive of capital, surplus value stands for the object, in which the social is converted into asocial. Marx targeted the asocial tendency of the drive of capital, when he described the capitalist production as "production for the sake of production" (Marx 1990: 742),[12] self-sufficient production without external, social purpose. To repeat, if surplus value is enjoyment of the capitalist system then this implies that the aim of capitalist production is expansion of uselessness and progressive dissolution of the social through the asocial. The primary aim of overproduction is the satisfaction of systemic drive for surplus value. If something like satisfaction of human needs and the preservation of life still takes place in this framework, it is merely accidental and in no respect follows from the tendencies of capital. In the end, the perseverance of capitalism in business as usual even in the epoch of accelerated climate breakdown and systemic destruction of living conditions only additionally demonstrates that the logical development of capitalism moves from the social to the asocial. Climate breakdown then appears as the ultimate surplus product of capitalism, the tantamount of its disastrous pursuit of uselessness in the guise of perpetual augmentation of profit and economic growth.

Ironically, and moreover anticipating the rejection of twentieth-century liberalism and neoliberalism which sought to legitimise its own assumption of *social* egoism in his work, Adam Smith intuited the asocial dimension of capitalism in general and in particular of private interest that according to him motivates human beings to sustaining social relations. Smith, for instance, draws attention to the fact that the private interest of corporations such as East India Company necessarily stands in open contradiction with the public interest (Smith 1999: 222–223). With this contradiction at hand the introduction of the notorious invisible hand does not strive to solve the problem by assuming an immanent tendency of the world market to equilibrium, but rather recurs to a metaphysical force, which supposedly guarantees the social embedding of the wealthy, the corporations and the market.[13] We can recall in passing that the invisible hand appears in Smith's work only a couple of times. A much more common concept is the metaphysically charged term Providence, which gives Smith's political economy a theological and teleological twist. At the same time, the invisible hand and Providence stand for the regulative force, which pushes for social embedding of asocial tendencies of the wealthy, of corporations and ultimately of capital. To translate it in Lacanian language, if the Market is the modern figure of the Other, the symbolic order, in which we are all immersed as political subjects and as social beings, then the invisible hand and Providence stand for "the Other of the Other", the metaphysical guarantee of the Other's completeness, stability and equilibrium, as well for its fundamentally social character. As soon as the hypothesis of the Other of the Other becomes inoperative, the Market turns out to be in permanent structural disequilibrium, and, what is more, a fundamentally asocial order, whose perseverance and reproduction ultimately implies death for human subjects. To claim "There is no such thing as society" or "Greed is good" means to make the step that was still unimaginable for Smith and fully embrace the asocial tendency of the capitalist drive for permanent economic growth and augmentation of value for the exclusive sake of its augmentation.

The populist negation of the reality of climate breakdown, or open indifference towards the unfolding of its consequences, provides another, most contemporary expression of the asocial tendencies of capital. With the ongoing transformation of neoliberal capitalism into an openly authoritarian order, contemporary populists show no trace of the outlived belief in capitalism as a *social* mode of production. In this regard, populist politicians such as Donald Trump, Jair Bolsonaro or Viktor Orbán, as well as entire political parties such as Alternative for Germany (AfD) or the US Republicans indeed serve as the ultimate exemplification of the asocial tendencies of capital. One often hears that the aim of populists is causing fear or anxiety. But is this really the case? Already a superficial glimpse shows that the aim of populism is causing resentment, an asocial affect *par excellence*. In contrast to anxiety, which in spite of all appearances can motivate the subject to social action, resentment aims at the other's repression and even annihilation, as the recent developments in the refugee crisis have shown. So, it is completely "logical" that populism enforces racism (annihilation of the ethnic other: refugees, migrants, Muslims, Jews, etc.), sexism (annihilation of the sexual other: homosexuals, transgender, women) and finally classism (annihilation of the economic other: proletarians, free-lancers, homeless, elderly, etc.).[14]

In the same move the populist denial of climate breakdown points out that another ideal, which grounded modernity, reached its limit, the ideal of "mastering nature" shared by modern science and capitalist economy. The idea of mastering – hence of exploitation – suggests that the alteration of environmental conditions of life and the progressive dissolution of the social are each other's flipsides: if climate breakdown confronts us with the truth of enforcement and implementation of the modern epistemo-economic ideal, it also exposes the realisation of the asocial dimension of capitalism. In this respect, climate breakdown is the ultimate surplus product of capitalism.[15]

The advocates of economic growth in our epoch of irreversible and accelerated climate breakdown ultimately admit that capitalism allows no social economy – only an asocial one, which continuous to dissolve the social relations and the environmental conditions of life. With regard to this capitalist tendency, Marx's abbreviation "M–M′" (money – increase of money) stands for significantly more than the self-valorisation of capital. If in capitalism economic exchange is considered the paradigmatic realisation of the social link – what Marx abbreviates as "C–M–C" (commodity – money – commodity) – then the subversion of the social dimension of exchange is already indicated in its capitalist inversion – "M–C–M′" (money – commodity – increase of money). What is indeed fictitious in "M–M′" is the social embedding of this presumable "automatic growth", the social dimension of surplus value. To repeat once more, Lacan's homology implies that surplus value is defined through uselessness and in this regard behaves like enjoyment according to Freud. In both contexts, the object of the drive is a surplus of the asocial over the social, the parasitism of asocial on the social. Both surpluses ultimately equal systemic waste, another type of surplus object, which already on the level of its phenomenology demonstrates nothing but uselessness, and to which we will therefore draw our attention in the following.

Wasteland of asociality

In his unpublished seminar *Les Non-Dupes Errent*, Lacan offers a minimal definition of modern society by pointing out how humans, in contrast to every other living being on Earth, produce tons of objects that end up as waste:

> It is altogether unique that everything humans create ends up as waste. A single thing which preserves a little dignity, are the ruins, but get out of your shells, all the same, a little bit, and you will see wrecked cars piling up, and you will notice that wherever you put your foot, you put your foot on something that – where people have tried in every possible way to recompress old rubbish in order not to be submerged by it, literally.
>
> (Lacan 1974, own translation)

Lacan's way of grasping the hidden kernel of modern society bears witness to Plato's famous allegory of the cave. Similar to Plato, Lacan describes an initial situation from which we are unable to suitably grasp this situation. As long as we are inside of "our shells", we do not notice what is actually going on. Only when we obtain a certain distance and "step out" we discover the truth.[16] However, the key difference between Plato and Lacan lies in their respective treatment of the "outside view" of the cave. In contrast to Plato's allegory, for Lacan, the outside does not refer to a world that exists independently of the cave, but to a world that is intrinsically shaped by the functioning of the inside. What we find as soon as we leave the cave is not the pristine world of "Mother Nature", but a profoundly, albeit strangely, human world. Instead of facing the rhythm of day and night or the reflection of the water, we see wrecks piled up into the sky. Wherever we look and wherever we go, we find ourselves in the middle of utterly human junk. Unlike Plato, the true discovery in Lacan's allegory is therefore not that there exists a "whole new world" outside of the cave waiting for us, but that the capitalist drive for permanent growth only works on the basis of the permanent growth of waste, and that the only way to keep going/growing is by excluding, or repressing, these unwanted leftovers from our cave.[17] Hence, there is no world of ideas or self-regulating nature outside the opaque reality of the cave, only the world of abjects, a world (*monde*) that is revolting (*immonde*) (Lacan 2013: 62).

In the words of Marxist geographer Neil Smith (2008: 7–8), one could say that Lacan's allegory bears witness of how "capital transforms the shape of the entire world. No God-given stone is left unturned, no original relation with nature unaltered, no living thing unaffected".[18] To substantiate this point, it is worth taking a closer look at Lacan's reference to the car cemetery, which can be better understood by pointing to the historical context in which his allegory is embedded. Teaching in 1974, one year after the oil crisis, Lacan saw himself confronted with a moment in the history of capitalism when, for the majority of Western counties, it became the first time widely recognised how much of the wealth of capitalist societies relies on fossil fuels. The oil crisis was a "shock", especially for the car industry, which was not only a key factor of the economy at that time, but also heavily relying on cheap access to oil. In the wake of the crisis, a vast amount of automobiles suddenly became useless, nobody wanted to use their car anymore because it was too expensive to drive it, and even less were interested in buying one. While the sudden surplus of useless cars became visible throughout the crisis, it is crucial to emphasise that Lacan does not consider this surplus as an exception to the "normal" functioning of society, but as its necessary side effect. Being a psychoanalyst, Lacan is well aware about the glaring difference between saying that a crisis creates a problem and saying that a crisis makes a problem visible. The crisis is not responsible for producing a surplus/waste that would otherwise not have been produced, but to make visible a surplus/waste that is always produced. Waste is not an anomaly of capitalism, but a surplus product, which functions as the revolting (*immonde*) logical flipside of surplus value.[19]

Waste and surplus value are two intertwined expressions of the asocial face of capitalism, because it is through them that we become aware that capitalism does not produce "for us", but only "for itself", that there is virtually no other reason for production than production itself.[20] As long as we solely dwell within the ideological setting of capitalism, this asocial face of overproduction remains hidden. In major cities, for instance, one can often find a highly professionalised economy and infrastructure built to support and sustain the fantasy that waste "magically" disappears from one day to the other. While incineration plants work day and night to decrease the size and volume of waste, the garbage collection used to work in the early morning hours, when most of us sleep, to reduce the chances for us of getting in contact with this dirty business. While the production of waste is a constantly marginalised phenomenon in capitalist societies, it takes a lot of *work* to keep it at bay, and to prevent us from being "submerged" by the insurmountable piles of trash that are (over)produced day by day in order to keep the system alive.

In this way, it becomes clear why there are "good reasons" from a capitalist standpoint to keep us staying inside of our shells. Eventually, the lesson of Lacan's allegory is that there is no world out there waiting for us to be discovered, that there is nothing but vast piles of useless waste and that this waste is, in fact, the very foundation on which our cave is built – indeed an inverted Platonism (as Deleuze would likely put it), in which it is the ontologically lowest that sustains the existence of the ontologically highest. There is no world of pristine nature because this world, whatever it may have looked like, is long gone, destroyed and devastated by the same forces that keep the capitalist machine running. "Stepping out of our shells" does not mean to discover anything new, but to see the waste that is repressed day by day in order to keep growing/going.[21] Such an approach therefore allows us to understand the end of the world not as some sort of big bang catastrophe that razes our world to the ground, but to conceive it as a gradual and catastrophic "surplus product" of capitalism itself, as the flipside of growth-oriented economy and the self-valorisation of capital. The end of the world is nothing that "just happens" without us being able to do anything against it, as it is often presented in apocalyptic imaginations of alien landings or meteorite impacts. It rather is a sort of by-product of the very functioning of capitalism itself. The apocalypse is immanently inscribed into the ways of how capitalism works.[22]

In the end, nothing brings the asocial character of capitalism more to the point than its destruction of the environment and its resistance to confront the climate breakdown as its ultimate "surplus product". An important measure of (a)sociality, in fact, reflects in the way we conceive the relation between nature and culture. Capitalist modernity implemented two contradictory yet equally wrong views of nature: on the one hand, nature as an ontologically stable order outside of culture, and on the other hand, the idea that the condition of the social resides in the "mastering of nature" (in other words, its exploitation). Nature is neither an uncorrupted harmonious order, a pure outside, nor is it an existential threat to culture, wild chaos, which threatens to undermine human civilisation.[23] If anything, nature is an organised disequilibrium that the capitalist social link progressively pushes in a direction, in which it will no longer sustain the necessary conditions for the preservation of (human) life. Precisely for this reason a key component of the social and the asocial is in the relation to nature. What is ultimately at stake in environmental destruction is the moment when the asocial surpasses the social, when the social is "submerged" by the asocial. The uncanny mixture of rising sea levels, melting glaciers, shifting climate zones, spreading parasites and tropical diseases, more intense or more frequent severe weather phenomena (from forest fires to floods, storms and droughts) and, of course, the massive increase of environmental refugees – all these are symptoms of the asocial face of capitalism demonstrating its superiority over the social. If there is one lesson to be learned from the way "disaster capitalism" works, it is that it does not mind destroying its own foundations, that it is, to a certain extent, even prepared to produce without humanity.

Conclusion

In the introduction to this chapter, we emphasised the powerful implementation of the apocalyptic imagination as crucial for understanding the status quo in many parts of the world today. The apocalypse is easy to imagine in the catastrophic times we live in. Moreover, the apocalypse is being imagined for us by the cultural and entertainment industry. All we need to do is visit a movie theatre, read the newest science fiction novel or open a newspaper. While the apocalyptic imagination has already started to become part of our everyday life, this imagination tends to perceive the apocalypse in a mystifying way, as a threat that confronts us with the possibility of social breakdown.

What is most problematic about the apocalyptic imagination is that it shows us an imaginary future world, in which the social bond between humans is no longer maintained and in which some kind of asocial predominates. In the course of this chapter, we have pursued the argument that the apocalyptic imagination, despite or perhaps even because of its imaginary productivity and future projection, represses a more discomforting truth – that the asocial forces of the apocalypse are an inherent part of the capitalist ordering of the social. The asocial is therefore nothing to be projected in some hypothetical future, it is not even something to be prevented; it rather immanently structures the way the capitalist social link functions (or rather dysfunctions) in the present. The asocial is that part of the catastrophe, which is progressively brought about by capitalism itself. What is asocial, and in a sense apocalyptic, about capitalism is the aspect, in which this social organisation of production slowly but surely makes human subjects redundant. Here, too, Marx already foresaw things when he described the self-valorizing value with the term "automatic subject". The latter stands for a "subject", who cut all ties with the human realm of subjectivation: language, labour and life.

Behind their spectacular scenarios the common imaginations of the apocalypse create the false impression of it being a future event rather than a process unfolding in the present and, with its very unfolding, erasing the very conditions of possibility for the future. The best prosopopoeia of the capitalist system could thus be the old punk exclamation: *No future!* And it is no coincidence that Johnny Rotten from The Sex Pistols is a professed fan of right-wing populism.

Notes

1 Given the changing nature of the situation we are writing about, it seems appropriate to add for contextualization that the opening lines of this chapter were written in April 2020 and that we are writing from a Western European perspective on the matter.

2 A classic scene from a zombie movie would be that the protagonist escapes an attack of the zombies with the help of a random stranger who suddenly appears out of nowhere. The stranger takes our protagonist to his hiding place and offers food and a place to sleep. Shortly after our protagonist starts to feel safe again, he or she realizes that things are not what they seem. Suddenly, the true intentions of the stranger are revealed and the real horror begins. The basic lesson to be learned from the zombie apocalypse is therefore not how to survive despite the zombie attacks, but how to survive despite the other humans. In the zombie apocalypse, the survivors learn that "their neighbor is for them not only a potential helper or sexual object, but also someone who tempts them to satisfy their aggressiveness on him [sic], to exploit his capacity for work without compensation, to use him sexually without his consent, to seize his possessions, to humiliate him, to cause him pain, to torture and to kill him" (Freud 1961: 111).

3 Bauman frames his line of thought by recurring to Freud's account of the antagonism between drives and culture: "Freud would present socially exerted coercion and the resulting limitation of individual freedoms as the very essence of civilization: civilization without coercion would be unthinkable, given the 'pleasure principle' (such as the urge to seek sexual gratification or the inborn inclination of humans to laziness), which would guide individual conduct towards the wasteland of asociality if it were not constrained, trimmed and counterbalanced by the power-assisted and authority-operated 'reality principle'". In our reading of Freud, we propose to correct the perspective: the conflict between the pleasure principle and the reality principle is interior to "culture", and, indeed, what is interesting in Freud is that he enables to detect the asocial tendencies within the social, that is, within capitalism.

4 The English translates this destiny of the drive (*Triebschicksal*) as "turning round upon the subject's own self" (Freud 2001: 126). What gets lost in this translation is the ambiguity of *gegen*, which primarily means "against", but can also signify "toward". Despite this ambivalence Freud's main concern is to draw attention on the contradiction between the drive's satisfaction and the self-preservation of the individual.

5 Sophie's son plays a prominent role in *Beyond the Pleasure Principle*, he is the child playing the *fort-da* (away-there) game in order to bridge his mother's absence. Freud eventually evokes Sophie's permanent *fort*.

6 Freud's *Civilisation and Its Discontents* makes an important step further in this controversial terrain. Its publication date (1930) again coincides with a destabilising social and economic reality, the great depression, in which the contradiction between the tendencies of capitalism and the self-preservation of

mankind once more appeared in all its brutality. Long-term structural processes such as economic crisis or climate emergency, which bring the contradiction between the reproduction of capitalism and the interests of mankind to an entirely new level, only increase the relevance of Freud as a thinker of crisis.

7 See, for instance, Marx (1990: 254, 342, 346, 377).

8 Recall Marx's remark from *German Ideology* (Marx 2000: 226–227), according to which bourgeois institutions consider themselves as natural, whereas the institutions of feudalism are denounced as artificial, that is, cultural.

9 One can mention the usual critique of consumerism here, although the restriction of surplus enjoyment to consumption would risk re-psychologising enjoyment and the drive, thus removing the systemic dimension of uselessness from the picture. Lacan's implicit thesis is that the organisation of enjoyment around useless surplus is what characterises the epoch of capitalism.

10 Freud's most systematic account of this issue is *Group Psychology and the Analysis of the Ego* (1955).

11 The formulation "enjoyment of the system" addresses the continuum between the subjective mode of enjoyment and the social mode of production, in other words, systemic enjoyment (enjoyment pertaining to the system) and individual enjoyment in and of capitalism.

12 In the same context, Marx speaks of "accumulation for the sake of accumulation", which only additionally pinpoints the asocial character of capitalism.

13 "The rich only select from the heap what is most precious and agreeable. They consume little more than the poor, and in spite of their natural selfishness and rapacity, though they mean only their own conveniency, though the sole end which they propose from the labours of all the thousands whom they employ, be the gratification of their own vain and insatiable desires, they divide with the poor the produce of all their improvements. They are led by an invisible hand to make nearly the same distribution of the necessaries of life, which would have been made, had the earth been divided into equal portions among all its inhabitants, and thus without intending it, without knowing it, advance the interest of the society, and afford means to the multiplication of the species" (Smith 2009: 215). It is quite peculiar to see Smith formulate here his own "they do this without knowing it" (*sie wissen es nicht, aber sie tun es*) (Marx 1990: 166–167). For Smith, the social is a compulsion in the life of the rich, which forces them to restrict, if not overcome, their egoism. Marx inverts this optimistic perspective: the compulsion stems from the asocial tendencies of capital itself and sabotages every attempt in social action from within.

14 To these three features a fourth one can be added, anti-science, rejection of critical knowledge concerning nature, society and subjectivity.

15 During the corona pandemic the true duty of every citizen was formulated in all clarity: die for the economy. The fact that this message came out of the mouth of populist leaders should not seduce us in believing that the liberal advocates of the status quo do not subscribe to it. Capitalism is ultimately grounded on imposed sacrifice. Lacan intuited this when he insisted that capitalism imposes renunciation of enjoyment as the condition for production of surplus enjoyment. The first figure of enjoyment that needs to be renounced in order for capitalism to live (in other words, to accumulate) is life itself.

16 This point is crucial for psychoanalysis more generally, as Joan Copjec (1994: 63) emphasises, when she states that it "is absolutely central to the psychoanalytic project...that the cause which must necessarily exist is never present in the field of consciousness that it effects".

17 One of the most suitable images to support Lacan's allegory is probably the Great Pacific Garbage Patch. This massive vortex of ever-growing plastic and floating trash located in the North Pacific Ocean is precisely what we witness

as soon as we "step out of our shell". In fact, someone dropped by a helicopter over this vortex would probably drown in it, he or she would "be submerged by it, literally".

18 In a certain way the difference between the two allegories in Plato and Lacan also resembles the shift from the standard image of "Mother Nature" to the Anthropocene. If we read the Anthropocene as another, albeit more radical, way of proving that nature cannot be understand independently of human activity, that humans are inscribed into the geomorphological constitution of the Earth, then Lacan indeed offers a suitable image to capture the anthropogenic state of the world.

19 Similarly one could say that rubbish dumps are the negative of tax havens, "heterotopies", from the perspective of which the capitalist social space must be looked at.

20 With regard to the difference between *aim* and *goal* that Lacan brings up in his 11th seminar during his discussion of the drive, we could say that capitalism does not seek to reach any particular goal, but is solely structured around its aim, which "is simply this return to the circuit" (Lacan 1998: 179).

21 That also explains why the encounter with waste is often an uncanny experience, because it "is something which is secretly familiar [*heimlich-heimisch*], which has undergone repression and then returned from it" (Freud 1981: 245).

22 Naomi Klein (2008) proposed the term "disaster capitalism", which fittingly describes the present stage of capitalist development (after merchant capitalism, industrial capitalism and financial capitalism).

23 Incidentally, this is where the modern view of nature intersects with the typical misogynous masculine fantasy of Woman: the Mother and the Whore.

References

Bauman, Z. (2007) *Consuming Life*. Cambridge: Polity Press.

Copjec, J. (1994) *Read My Desire: Lacan against the Historicists*. Cambridge and London: MIT Press.

Freud, S. (1955) *The Standard Edition of the Complete Psychological Works of Sigmund Freud, Volume 18, Beyond the Pleasure Principle, Group Psychology, and Other Works*. London: The Hogarth Press.

Freud, S. (1961) *The Standard Edition of the Complete Psychological Works of Sigmund Freud, Volume 21, The Future of an Illusion, Civilization and Its Discontents, and Other Works*. London: The Hogarth Press.

Freud, S. (1981) *The Standard Edition of the Complete Psychological Works of Sigmund Freud, Volume 17, Infantile Neurosis and Other Works*. London: The Hogarth Press.

Freud, S. (2000) *Triebe und Triebschicksale. In: Studienausgabe, Volume 3*. Frankfurt: Fischer.

Freud, S. (2001) *The Standard Edition of the Complete Psychological Works of Sigmund Freud, Volume 14*. London: Vintage Classics.

Jameson, F. (2003) Future City. *New Left Review* 21, 65–79.

Klein, N. (2008) *The Shock Doctrine. The Rise of Disaster Capitalism*. London: Penguin.

Lacan, J. (1974) *Seminar, Book XXI, Les Non-Dupes Errent*. Unpublished Manuscript.

Lacan, J. (1998) *Seminar, Book IX, The Four Fundamental Concepts of Psychoanalysis*. New York: Norton.

Lacan, J. (1999) *Seminar, Book XX, Encore.* New York: Norton.
Lacan, J. (2013) *Triumph of Religion.* Cambridge: Polity Press.
Marx, K. (1990) *Capital, Volume 1.* London: Penguin.
Marx, K. (2000) *Selected Writings.* Oxford and New York: Oxford University Press.
Smith, A. (1999) *The Wealth of Nations, Books IV–V.* London: Penguin.
Smith, A. (2009) *Theory of Moral Sentiments.* London: Penguin.
Smith, N. (2008) *Uneven Development: Nature, Capital, and the Production of Space.* Athens and London: The University of Georgia Press.
Zupančič, A. (2006) When Surplus Enjoyment Meets Surplus Value. In: Clemens, Justin and Russell Grigg (Eds.) *Jacques Lacan and the Other Side of Psychoanalysis.* Durham and London: Duke University Press, 155–178.

9 The self(ie) in the Anthropocene

Doug Specht and Cat Snyder

Climate change is now fully back on the political agenda. 2019–2021 saw unprecedented fires across four continents; extreme weather events have also seen droughts in the USA, flooding in the UK and Pakistan, intensive air pollution in China and India, and ever larger and destructive typhoons over Taiwan and the Philippines (Pidcock et al., 2019). The rise of Extinction Rebellion, and the School Climate Strike Movement, led by Greta Thunberg, which alongside other long-standing campaigns, have brought issues of climate to the public consciousness, bringing about a sense of urgency for action. Even major financial institutions are now warning that climate change may affect investments (Dalby, 2013; Greenfield and Watts, 2020). It is also becoming ever clearer that these changes to the climate, the Earth's biochemical cycles and its biophysical parameters are the result of human activity (Dalby, 2014; Lewis and Maslin, 2015; Wright et al., 2018). Such is this level of change that scientists now widely agree that the world has embarked on a new ecological era, the Anthropocene (Dalby, 2014). The global scale of the change will leave a stratigraphic record for millions of years, denoting the shift to a time in which nature is no longer disconnected or external from human activity (Dalby, 2013; Knight and Harrison, 2013; Lewis and Maslin, 2015). While a start date for the Anthropocene is still being discussed, the effects of the new epoch are being felt globally, and the conditions of global capitalism that created them are being questioned by an ever-increasing number of people (Dalby, 2013; Greenfield and Watts, 2020).

Beyond simply providing a new framework for understanding human responsibility in climate change (Leinfelder, 2013), the Anthropocene adds an urgency and call for accountability from governments and institutions (Pattberg and Zelli, 2016). Local communities are increasingly confronting the physical impacts of the Anthropocene, and the argument that these impacts are created by natural causes – a notion designed to enable the rich and powerful to avoid their responsibilities – is being eroded (Dalby, 2013). Opposition is growing from NGOs, grassroots activists, charities, indigenous communities and individuals (Wright et al., 2018). At the same time, protests and civil disobedience are becoming more widespread (Della Porta and Parks, 2014). For many the Anthropocene signals the end of global

DOI: 10.4324/9781003128854-10

capitalist civilisation, that life will become increasingly precarious (Wright et al., 2018), and that we must now move towards no longer seeking to save the planet, but instead we must learn how to die in order to allow for humanity to survive (Scranton, 2015).

Against this bleak backdrop, protests around climate change are beset with smiling people, taking selfies and presenting themselves online. And while it has been shown that selfies can form part of a participation of social movements (Liu et al., 2017), new questions are worth asking as to their role during protests in the death throes of civilisation. Are these selfies the last moments of capitalism clinging on through the commodification of the protester? Are they a narcissistic undertaking of performative politics? Are they a demonstration of participating in the human? Or are they one last chance to see ourselves before the world turns black at the end of time?

Selfies have become a ubiquitous part of society, and this has long been part of the online culture, with #Selfie appearing as early as 2004 on Flickr and MySpace (Rutledge, 2013) – the concept of the selfie bled into all areas of society in 2013 when it beat 'twerk' to become the Oxford Dictionaries word of the year (Ryan, 2013; Gibbs et al., 2014). This choice was prompted by the 17,000% rise in the use of the word in the 12 months leading to its selection (Oxford Dictionaries, 2013). In the year following, Google statistics suggest that 93 million selfies were being taken every day on Android platforms alone (Brandt, 2014). In 2017, the Saatchi Gallery made them the focus of their exhibition in conjunction with tech giant Huawei (Saatchi, 2017).

Despite these huge numbers, what a selfie is, and what purpose it serves remain a source of contention. Senft and Baym (2015) suggest they 'remain fundamentally ambiguous, fraught and caught in a stubborn and morally loaded hype cycle' (p.1588). At their simplest, the selfie is defined as a photographic self-portrait and one that is principally taken with a smartphone or webcam and shared via social media (Bealing, 2016; Diefenbach and Christoforakos, 2017). This definition rather implies that they are a new phenomenon, but it is also important to recognise that while their proliferation is modern, people have been taking selfies since the invention of the camera (Cornelius, 1839). The act of taking a selfie is often seen as something frivolous (Davis, 2014), which leads to fret, social consternation and low self-esteem (Weiser, 2015; Humphreys, 2018). Conversely, they are also seen by many as being born from low self-esteem, or narcissistic traits (Griffiths et al., 2014; McCain et al., 2018). These ideas though might be seen as overly reductionist, and the pervasiveness of selfies, and selfie takers, in everyday life suggests that a more nuanced and deep examination of their role in society is needed (Murphy, 2015; Diefenbach and Christoforakos, 2017), and particularly within the context of the self in the Anthropocene.

In order to undertake this deeper examination, this chapter follows Senft and Baym's (2015) notion that a selfie is both an object and a practice which seeks to fulfil the role of transmitting human feeling, and to send messages to individuals, communities and audiences. These acts and objects form an important part of presencing, creating a public presence beyond the bodily presence (Couldry, 2012), a process that Humphreys (2018) notes is especially important as we engage in contemporary social rituals of loss. In this chapter, we reflect on the way in which selfies are used to create presence at the end of time, how political intent has crept into the process of selfie taking (Pisani, 2015) and how selfies are used to create representations of ourselves to be consumed (Humphreys, 2018) in the face of ecological melt-down. As the USA pulls from a Paris agreement and Brazilian policy deliberately denies the rainforest conservation crisis, the Anthropocene becomes less unintentional impact of human activity with deliberate denial of impending doom. In the face of environmental crisis and the increasingly apocalyptical political climate of neglect, what place do selfies hold, if any at all?

The narcissism of the selfie

Writers such as Giroux (2015) speak of selfie culture as epitomising the disintegration of the public sphere that sustains democratic society, and the popular press, at times supported by academic research, is keen to point to selfies as a narcissistic undertaking that merely seek instant gratification and self-satisfaction (Halupka, 2014). The image points directly at the user. The focal point being the user at close range, posted by the user themselves, creating a 'looped gaze in which the photographer and photographed are conflated subject positions' (Brager, 2015). The taker is not only the producer but also the subject and the publisher all in one, adding to the conflated sense of self and performance (Brager, 2015; Frosh, 2015).

Selfies are commonly seen as being born from a narcissistic necessity (Best, 2016; Moon et al., 2016) directly feeding the narcissist's need to gain attention and to maintain an inflated sense of self (Morf and Rhodewalt, 2001; McCain et al., 2016). Along with the addictive nature of social media (Griffiths et al., 2014; Andreassen, 2015), this can develop into performative and narcissistic uses of social platforms, where selfies are used to boast and put forward the best possible version of oneself for followers to see. Technology also allows for instant modification of images through editing and filters, further adding to the ability to instantaneously present an altered and often polished version of reality. Popular culture and celebrity influencers social media accounts are full of unrealistic ideals presented through photoshopped images, encouraging a competitive and fabricated perception of beauty (Shaw and Lee, 2012). Selfies are then ripe for being used for self-enhancement and modification.

Beyond promoting unrealistic perceptions of self, more significant ramifications exist on both individual and societal levels. Han (2018) even equates selfies with self-harm, suggesting that to escape emptiness, one 'reaches for the razorblade or the smartphone' (p. 25). Selfie culture has become a place where getting the perfect image is desired above all, even if it includes putting yourself in danger. The attempted suicide of Danny

Bowman (Molloy, 2014) who was purportedly addicted to selfies, and the 259 selfie-related deaths between 2011 and 2017 (Bansal et al., 2018) provide sobering examples of the potentially fatal side effects of performative selfie culture.

Attention-seeking narcissism and desire for feedback from followers correlates with frequency of social media use (Weiser, 2015; McCain and Campbell, 2018; Shane-Simpson et al., 2020). The often ego-enhancing environment of social media only mirrors the narcissism of the very authoritarian leaders who reject the notion of the humanity's complicit role in the coming doom of the end of times. Although evidence is scant to show the connection to selfie usage and narcissism, the themes of self-presentation exist for those who already present with narcissism (Barry et al., 2017). And while few direct links have been shown, there is much public discourse around selfies, social media and how these are connected to low self-esteem, narcissism, depression and loneliness (Hawi and Samaha, 2017; McCain et al., 2016). Getting the perfect selfie has, for some, become obsession. While narcissism, mental health and suicide certainly exist independently from selfies and social media, they provide an avenue for and exhibition space to potentially amplify these issues.

Selfies can create an imaginary sense of connectedness, which can be seen when politicians and celebrities get wise to the fact that selfies create an instant intimacy with their audiences by posting a close-up photograph (Ekman and Widholm, 2017). Selfies often reduce society to a collection of nomadic individuals fuelled by celebrity culture and false notions of community (Giroux, 2015), what Honneth (2009) calls the 'abyss of failed sociality' (188). It is only within a collective action, hashtagged to the appropriate cause, that a selfie can have some impact.

Outside of a collective voice, a selfie lacks the power to impact meaningful change and is simply a photograph at close range of the person who took the photo. Perhaps a selfie has positive elements of self-expression or social interaction (Lee et al., 2015) or event marking (Kuntsman, 2017), given the urgency of climate change, the time spent taking, potentially editing, posting a selfie, and obsessing over the likes and views would be better used engaging in political and potentially life-changing action.

Selfies as an expression of existence

There is then an unquestionable level of fear around the effect of selfies, much of which extends from this literature on narcissism. Indeed, some places, such as parts of India, have banned selfies or set up 'no-selfie zones', ostensibly for safety reasons, but also feeding the narrative that they are somehow harmful for the individual (Lazzaro, 2016). These narratives though do not consider that there are multiple kinds of selfie practice (Humphreys, 2018); while some evidence suggests that internet use and portraying yourself digitally can increase loneliness (see Moody, 2001), there is also evidence that these acts can decrease loneliness (see Fokkema and Knipscheer, 2007). Indeed, scholars such as Griffith and Papacharissi (2010) have noted that while vloggers mostly talk about themselves, something that many would call narcissistic (Bealing, 2016), the act of vlogging itself is more about creating an audience, or making human connections. The desire to make such connections then may well also be the root of many selfies. Jessie Fox of the Ohio State University School of Communication has suggested that narcissism isn't just about being an egomaniac, but is instead driven by an underlying insecurity (cited in Murphy, 2015). Within the context of climate change and the Anthropocene, we might then ask, what is more insecure than our future? de Beauvoir also noted that what often appears to be narcissistic behaviour in women is in fact a response to the limitations and insecurity of femininity. Selfies then are perhaps less of an indication of moral decline, but rather are an important part of technology-enabled self-exploration and actualisation (Rutledge, 2013).

Selfies consist of more than 'duck faces' (Senft and Baym, 2015), and, as noted above, there is scant evidence of links between selfies and mental health (*ibid.*). Instead, selfies can be seen as an important part of self-expression and social interaction (Lee et al., 2015). Humans have long shown a need for self-exploration, with self-study and self-observation being an important part of exploring identity (Rutledge, 2013). Russell et al. (2015) in their examination of the online hashtag #HospitalGlam found that many of those posting under this hashtag felt that they were reasserting agency over their own bodies in an otherwise uncertain space. In this respect, selfies then might help us to figure out who we are, and what we are (Diefenbach and Christoforakos, 2017), and with research suggesting that people are most likely to engage in these activities when anxious (Murphy, 2015), being made visible through a selfie can become a powerful way of positioning oneself in time and space, especially in moments of distress (Kuntsman, 2017). This notion is explored in depth by Gibbs et al. (2014) in their study of selfies at funerals. While the act is seen as abhorrent to many, Gibbs et al. noted that selfies are in many ways public outpourings of grief which fill a void left by waning institutionalised mourning rituals in Western society. Again, within the context of climate change protesting, in which themes of grieving for the planet are apparent, we might understand the selfie as a moment of public mourning. Even, staunch selfie critic, Giroux (2015) concedes that there are many instances in which people are using selfies in a proactive way that does not buy into mainstream selfie culture.

The way in which selfies are being used as a way of both exploring ourselves and expressing grief may well go some way to explaining why we are increasingly seeing the political use of selfies (Kuntsman, 2017) as a form of online expression (Halupka, 2014). While these actions have been dismissed by scholars such as Morozov (2009) as clicktivism, an online action that is so easy as to be dismissed as frivolous (see Karpf, 2010), and which is disconnected from democracy (Halupka, 2014), selfies within and about protests play a range of other important roles. They present an alternative to the mainstream media (Aziz, 2017), and provide a proof of an event witnessed, by combining a presentation of self along with the event (Koliska and Roberts, 2015). Political selfies are no longer a singular genre, but are now an area that encompasses a wide range of ideologies (Pisani, 2015). In all contexts though they also fulfil the role of self-staging, a way of communicating and presenting an intended image to others (Diefenbach and Christoforkos, 2017; Humphreys, 2018).

Selfies then can go beyond a mere demonstration of interest and support of a movement (Halupka, 2014), but instead can provide a space for self-reflection, and moreover the selfie constructs a new political space, and reaffirms the existence of the protester as they wrestle with their own minimisation at the hands of authoritarian states (Nunes, 2017). The selfie becomes a political object through the takers' engagement with the cause (Halupka, 2014), and can become a means of claiming 'speakability, agency and survivor-hood in the face of violent silencing and stigmatization' (Kuntsman, 2017: 17). The taker of the selfie is reaffirmed through their existence both as an embodied presence before the lens, and as a digital image distributed across social networks (Nunes, 2017).

Yet, while there is a clear case for the importance of reflective, performative and reaffirming nature of the selfie as part of the process of creating spaces for citizenship (*see* Frosh, 2015), it is important to note that they are not a 'fix-all' and work on selfies and the feminist movement has demonstrated well the entanglements which might ensue. It is clear that not all selfies are feminist, as Giroux (2015) bluntly puts it, 'selfies aren't empowering, they're a high tech reflection of the fucked-up way society teaches women that their most important quality is their physical attractiveness' (161), but it is also important to recognised that selfies may play multiple roles, and that they may also allow women to play with identity representation, one that is reminiscent of the1970's Feminist Art Movement's use of Polaroids (Pisani, 2015). The message is clearly that selfies are more complex than just being frivolous narcissistic pictures of how much better our lives are than someone else's. Critics such as feminist artist Trudy comment that the self-expression of selfies allows for the marginalised to self-author their stories in the face of misogyny. 'They are records of my own existence … on my own terms' (Trudy, 2015).

An important way of looking at selfies then is to remember that stripping an image or painting of its context of production will ignore the politics of that production (McGill, 2018). Just because we might now be able to log into social media to quickly post images of ourselves does not diminish the act of reaffirmation of both ourselves and our social bonds (Knowles et al., 2015). Especially when looking at protests and within the context of the Anthropocene, it is important that we do not disassociate our understanding of selfies from this politics of production. Instead, we must study them as a quotidian experience and as a way in which we situate and connect ourselves with the world around us (Humphreys, 2018). Selfies form part of an intricate presentation of ourselves, and because of this they will always be a complex, and at times contradictory, practice (Diefenbach and Christoforakos, 2017), that may be seen as self-serving or edging towards the narcissistic (Weiser et al., 2015), but ultimately are about presencing and providing visual evidence that we exist (Humphreys, 2018).

It is clear then that the selfie is an increasingly important field of study, and especially within the Anthropocene where their creation and use become multifaceted and in which they are forced to contend with neoliberalism, narcissistic authoritarianism and fears around the end of time. Within this context, they appear to play an important role in the process of emplacement and affirmation of self. Furthermore, the authenticity provided by the documenting of participation within a cause helps disparate peoples recognise themselves in one another (Strauss, 2003), an important part of building global networks of resistance. The challenge here, following the work of Nunes (2017), is to see selfies as an act of placemaking even when they express themselves in seemingly out-of-places poses and hashtags.

In order to explore the contradictions between the way in which selfies are perceived as narcissistic and the way in which they are used as presencing device within the context of climate change, a large-scale analysis of online selfies was undertaken. Working from the two conflicting hypotheses, which were born of the authors' conflicting views on the subject, the aim was to establish, to some extent, whether selfies within the climate change movement were being presented as a self-serving act of virtue signalling, or if they might indeed be suggesting an acknowledgement of existence within the context of a dying planet and politicians who are apparently deaf to these concerns.

Accessing the Harvard DataVerse 39,622,026 tweet IDs were downloaded. These were related to #ClimateChange and were collected between September 21, 2017 and May 17, 2019 using the Twitter API and Social Feed Manager. There is a noted gap in data collection between January 7, 2019 and April 17, 2019 (Littman and Wrubel, 2019). Extraction of detailed information from IDs was undertaken using Hydrator.[1] Following this process, which could not rehydrate deleted posts, 27,426,202 tweets were available for analysis. The database was paired down further through the use of hashtags related to protest marches. These included *#Climatemarch*, *#climatestrike*, *#climateaction* and *#schoolstrike4climate*. Of these only those tweets with an image were analysed. Once other tweets were removed 270,525 remained for the analysis.

To begin the analysis, it was first required to isolate tweets by individuals rather than bots or organisations. This was undertaken by removing all instances where an account had posted the same image repeatedly. The remaining tweets were then modelled in Gephi to help explore linkages and networks, and individual photographs were analysed for their content. Images were sorted into those that could be defined as a true selfie (taken by the person in the photo), group selfies, 'someone else's', where the image was set up to look like a selfie, but was taken by somebody else. And finally, images that were not selfies.

Selfies as resistance

From the data available, it quickly became clear that the self-conscious idea of posting a selfie for the sake of being a selfie was not something that was a feature of the climate change data. From the full set of near 40 million tweets, just 76 used the hashtag *#selfie*. However, from the subset of tweets created using our hashtag searches it was clear to see that a good deal of selfies existed – although overall they only accounted for a small number of the images due to the large number of images being posted by organisational accounts as ways of sharing information, a result of the low number of characters allowed on Twitter.

Research suggests that using social media to post selfies relates to a need to belong (Knowles et al., 2015), that they are attention seeking (Sung et al., 2016), that they are used for self-promotion, and demonstrating to others heightened skills, beauty or intelligence (Diefenbach and Christoforakos, 2017). But also that they can serve as a seeking tool, a desire for acceptance and a wanting to belong, a way of escaping loneliness and disconnection (Knowles et al., 2015). Indeed, other research has suggested that those who post selfies are more likely to befriend strangers, and this could be seen as an extension of wanting to be part of an 'in crowd' (ibid.).

To this end, the tweets collected for this chapter were modelled into networks using Gephi, running a modulatory analysis revealed 3,990 groups within the network, and generally weak ties (see Figure 9.1). There was also a correlation between those who had few or no ties and the posting of selfies. While this is partly attributable to those with many ties being corporations and NGOs who were mostly sharing more corporate images, it also supports the idea that those with less ties are more likely to post selfies. What becomes clear very quickly is that selfies are predominantly being taken by those with weak ties to the group. Many of the selfies presented in the data have no connections to other nodes in the network, and the majority just have one. This suggests that the use of selfies in these protests is part of an aim to show belonging in part of a network in which the individual is rather distant, or transient. This need is related to the 'pics or it didn't happen' sentiment (Murphy, 2015), but also the need to construct an identity in which the individual appears to be actively engaged in improving the world, challenging injustice and creating a positive sense of impression management (Pisani, 2015; Sung et al., 2016; Diefenbach and Christoforkos, 2017; Humphreys, 2018).

The nature of these protests though adds an additional layer to these selfies. And looking at individual images it becomes clear that the notion of virtue signalling is far from the only motivation in these moments. Indeed, there is a sense of a very real movement and political resistance within these

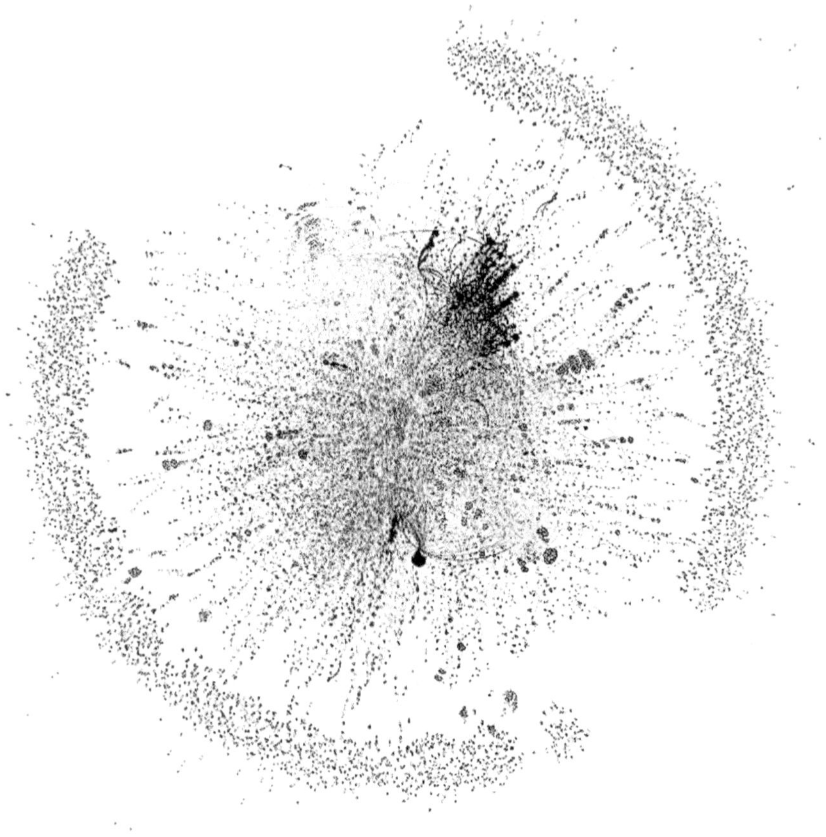

Figure 9.1 Networks of those tweeting about climate change. Selfies were most likely to be posted by those in the outer circle, with no ties to the rest of the network.

selfies. This is something that has been seen previously in the feminist cultural movements, as well as in political resistance in Turkey. The selfie is not a representation of the self *at* a protest, the selfie *is* the protest (Pisani, 2015). As Bealing (2016) notes, one's image in the form of a selfie may well reveal more about one's inner self, than of their outwardly projecting image.

While selfies are often dismissed because they document the everyday, turning it into a ritualised performance (Humphreys, 2018), there is less ritualisation in these climate marches, which for the data used were still sporadic. Rather in these moments the selfie is communicating much more than 'I am here', but is seeking to build relationships through the reactions of others to the selfie (Sung et al., 2016). The climate crisis is one that is very much beyond the control of the individual; no one person can help solve or change the onward march towards apocalypse, but the selfie offers a sense of control, a control over your own narrative and personhood, providing an authority and autonomy which has been stripped away by the inevitability of climate change (Murphy, 2015; McGill, 2018). While the Anthropocene points to human consequences on the environments, selfies provide a small window of agency for redemption. But more than this, the social media selfie is a natural avenue for grief, with the medium now a ubiquitous part of everyday life (Gibbs et al., 2014). This autonomy, and grieving for the planet, coupled with a very real sense of being ignored by governments and excluded from the processes of change only serves to speer efforts to reconnect with others through the use of selfies and other computer-mediated means (Baumeister and Leary, 1995; Knowles et al., 2015).

Much research looks to how women have used selfies to reassert themselves in times of eroding rights. McGill (2018) makes the observation that, without the ability to look, and to have that look acknowledged, expressed, represented, women in culture can never be subjects, only objects. In that instance, and without overstating the liberation potential of digital technologies, it is possible to propose that, in the context of the everyday, selfie practices offer a chink in the armour of male gaze dominance, a crack where the light of the female gaze seeps through. As both subjects, photographer and publisher, the selfie gives powerful agency to allow control over portrayal of the female form which historically was left to the male and powerful (Routh, 2016). If the selfie holds a particular fascination for women, it only takes the briefest glance at the trajectory of Western history to begin to understand why this might be the case. Despite advances towards liberation and a more inclusive culture, visual representation remains, as Griselda Pollock (2003) puts it, a site of privilege. This is also seen in one of the major feminist selfie archives on Twitter which was initiated in the summer of 2014, when Turkish Deputy Prime Minister, Bulent Arinc, said in a speech that women 'should know what is decent and what is not decent. She should not laugh loudly in front of all the world and should preserve her decency at all times' (O'Connor, 2014). The response was a barrage of selfies of women laughing posted to social media sites. Images from both Turkish women and other women internationally joining in solidarity were posted a short time after the speech was made public, including a contribution from Emma Watson, the United Nations' Goodwill Ambassador/Feminist It Girl since July 2014 (Pisani, 2015).

While in some respects these confirm the self-presentational values of selfies, it also demonstrates that this is only part of the story and that there is much more around the construction of the selfie that provides a bigger narrative (Diefenbach and Christoforakos, 2017). The term 'narcissism' is often used in describing selfie takers, but cultural theorist Imogen Tyler asserts that 'the accusation of narcissism was mobilized in the 1970s to denigrate identity politics and forms of sexual behavior deemed counter-normative…' (Pisani, 2015). It could be argued that the banning of selfies in parts of the world that are deemed authoritarian, and the dismissing of them as narcissistic is to dismiss protest and challenges to the status quo that come with the very real place and person making effects of selfies. To understand, then, the selfie in protest marches requires a shift away from the notions or personality traits and towards the motivations for their use (Wang et al., 2015).

Even if the motivation inspiring selfie use within protest is narcissistic or borne from performative intentions, the selfie holds a powerful place in challenging power, especially within the context of climate change. Shot from close range, selfies' sharp aesthetic confronts and even demands the viewer's attention. Since 2019, Australia's drought and temperatures have reached record highs contributing to catastrophic destruction of the entire continent with preliminary research and commentary pointing to anthropogenic climate change as the cause. As Australia burned, it put a spotlight yet again on the urgency of climate change and complicit nature of those who hold power to change it.

Conclusions

Ryan (2013) made the satirical quip that when posting a selfie, what the poster is really saying is 'Hey guys, I'm by myself [...] can you please somehow indicate that other humans are out there so that I do not collapse into my own loneliness' (n.p.). However, this chapter suggests that this cry for the other is a part of the selfie process in times of distress, particularly when faced with monumentally large issues such as climate change. Selfies, even posted in isolation, can join a larger community, and more connected networks, and add to its strength of the collective call to action. Those who might be on the periphery of civic engagement can participate.

Selfies taken at the climate marches are then not expressions of pride, or narcissism, or even the sharing of the mundane and meaningless, but are calls for affirmation (Ryan, 2013; Humphreys, 2018). Popular media have reinforced the notion that selfies are frivolous and narcissistic, discounting their complexities, and seeing to eradicate their political power (Pisani, 2015; Abidin, 2016). Yet to discount them as meaningless ignores this power of the selfie. Looking straight into the camera lens and holding the powerful to accountability is a deeply meaningful expression. Using one's agency to create, produce and disseminate a self-portrait, and participate in a larger cause is truly impactful.

The Anthropocene is a product of capitalist economic relations, started by fossil fuel-led industrialisation, and driven by the processes of forcing capitalism into the dominant economic model for human development. Within this the selfie might be seen by many as another example of capitalist urges, but as the relationship between humans and the earth becomes more distant, threatening our very existence, the selfie is the natural capitalist response (Wright et al., 2018). They are, though, not only the linking of socio-technical systems: in the fight against climate change and ecological breakdown they have become a place for reflection and a social act (Humphreys, 2018). An important tool in the process of saying 'I exist' and 'I don't want to die'. As Murphy (2015) reminds us, selfies are not just another form of communication. A picture is worth a thousand words!

Note

1 https://github.com/DocNow/hydrator.

References

Abidin, C. (2016). "Aren't these just young, rich women doing vain things online?" Influencer selfies as subversive frivolity. *Social Media+ Society*, *2*(2), 2056305116641342.

Andreassen, C. S. (2015). Online social network site addiction: A comprehensive review. *Current Addiction Reports*, *2*, 175–184.

Aziz, F. (2017). Performing citizenship: Freedom march selfies by Pakistani Instagrammers. In: A. Kuntsman (ed.) *Selfie citizenship*, 21–28. Cham: Palgrave Macmillan.

Bansal, A., Garg, C., Pakhare, A., & Gupta, S. (2018). Selfies: A boon or bane? *Journal of Family Medicine and Primary Care*, *7*(4), 828.

Barry, C. T., Doucette, H., Loflin, D. C., Rivera-Hudson, N., & Herrington, L. L. (2017). "Let me take a selfie": Associations between self-photography, narcissism, and self-esteem. *Psychology of Popular Media Culture*, *6*(1), 48.

Baumeister, R. F., & Leary, M. R. (1995). The need to belong: Desire for interpersonal attachments as a fundamental human motivation. *Psychological Bulletin*, *117*(3), 497.

Bealing, L. (2016). The fairest of them all: Selfie culture, gendered vanity, and art. [Online] Available from: https://lucywriteswords.wordpress.com/2016/09/21/the-fairest-of-them-all-selfie-culture-gendered-vanity-and-art/ [Accessed December 2019].

Best, C. M. (2016). 'These snaps are made for talking': Visual communication and instant expression on snapchat. *AoIR Selected Papers of Internet Research*. 6.

Brager, J. (2015). Selfies| The selfie and the other: Consuming viral tragedy and social media (after) lives. *International Journal of Communication*, *9*(12), 1660–1671.

Brandt, R. (2014). Google divulges numbers at I/O: 20 billion texts, 93 million selfies and more. [Online] Available from: www.bizjournals.com/sanjose/news/2014/06/25/google-divulges-numbers-at-i-o-20-billion-texts-93.html

Cornelius, R. (1839). The first light picture ever taken. [Online] Available from: https://publicdomainreview.org/collection/robert-cornelius-self-portrait-the-first-ever-selfie-1839 (Accessed February 2020).

Couldry, N. (2012). *Media, society, world: Social theory and digital media practice*. Cambridge and New York: Polity.

Dalby, S. (2013). The geopolitics of climate change. *Political Geography*, *37*, 38–47.

Dalby, S. (2014). Rethinking geopolitics: Climate security in the Anthropocene. *Global Policy*, *5*(1), 1–9.

Davis, B. (2014). Ways of seeing Instagram. *ArtNet* [Online] Available from: https://news.artnet.com/exhibitions/ways-of-seeing-instagram-37635 [Accessed December 2019].

Della Porta, D., & Parks, L. (2014). Framing processes in the climate movement: From climate change to climate justice. In: M. Dietz and H. Garrelts (eds.) *Routledge handbook of the climate change movement*, pp. 19–30. Abingdon: Routledge.

Diefenbach, S., & Christoforakos, L. (2017). The selfie paradox: Nobody seems to like them yet everyone has reasons to take them. An exploration of psychological functions of selfies in self-presentation. *Frontiers in Psychology*, *8*, 7.

Ekman, M., & Widholm, A. (2017). Political communication in an age of visual connectivity: Exploring Instagram practices among Swedish politicians. *Northern lights: Film & Media Studies Yearbook*, *15*(1) 15–32.

Fokkema, T., & Knipscheer, K. (2007). Escape loneliness by going digital: A quantitative and qualitative evaluation of a Dutch experiment in using ECT to overcome loneliness among older adults. *Aging & Mental Health*, *11*, 5.

Frosh, P. (2015). Selfies| The gestural image: The selfie, photography theory, and kinesthetic sociability. *International journal of communication*, *9*(22), 1670–1628.

Gibbs, M., Nansen, B., Carter, M., & Kohn, T. (2014). Selfies at funerals: remediating rituals of mourning. *AoIR Selected Papers of Internet Research.*

Giroux, H. A. (2015). Selfie culture in the age of corporate and state surveillance. *Third Text*, *29*(3), 155–164.

Greenfield, P., & Watts, J. (2020). JP Morgan economists warn climate crisis is threat to human race. *The Guardian.* [Online] Available from: www.theguardian.com/environment/2020/feb/21/jp-morgan-economists-warn-climate-crisis-threat-human-race (Accessed February 2020).

Griffith, M., & Papacharissi, Z. (2010). Looking for you: An analysis of video blogs. *First Monday*, *15*(1) doi: https://doi.org/10.5210/fm.v15i1.2769.

Griffiths, M. D., Kuss, D. J., & Demetrovics, Z. (2014). Social networking addiction: An overview of preliminary findings. In: K. P. Rosenberg and L. C. Feder (eds.) *Behavioral addictions: Criteria, evidence, and treatment*, pp. 119–141. London: Academic Press.

Halupka, M. (2014). Clicktivism: A systematic heuristic. *Policy & Internet*, *6*(2), 115–132.

Han, B. C. (2018). *The expulsion of the other: Society, perception and communication today.* Oxford: John Wiley & Sons.

Hawi, Nazir S., & Samaha, M. (2017). The relations among social media addiction, self-esteem, and life satisfaction in university students. *Social Science Computer Review*, 35(5), 576–586.

Honneth, A. (2009). *Pathologies of reason: On the legacy of critical theory* (Vol. 23). New York, NJ: Columbia University Press.

Humphreys, L. (2018). *The qualified self: Social media and the accounting of everyday life.* Cambridge, MA: MIT Press.

Karpf, D. (2010). Online political mobilization from the advocacy group's perspective: Looking beyond clicktivism. *Policy & Internet*, *2*(4), 7–41.

Knight, J., & Harrison, S. (2013). The impacts of climate change on terrestrial Earth surface systems. *Nature Climate Change*, *3*(1), 24–29.

Knowles, M. L., Haycock, N., & Shaikh, I. (2015). Does Facebook magnify or mitigate threats to belonging? *Social Psychology*, *46*(6), 313–324.

Koliska, M., & Roberts, J. (2015). Selfies| selfies: Witnessing and participatory journalism with a point of view. *International Journal of Communication*, *9*, 14.

Kuntsman, A. (ed.) (2017). *Selfie citizenship.* Cham: Palgrave Macmillan.

Lazzaro, S. (2016). This viral image makes a bold statement in defense of selfie culture. *Observer.* [Online] Available from: https://observer.com/2016/01/this-viral-image-makes-a-bold-statement-in-defense-of-selfie-culture/ [Accessed December 2019].

Lee, E., Lee, J. A., Moon, J. H., & Sung, Y. (2015). Pictures speak louder than words: Motivations for using Instagram. *Cyberpsychology, Behavior, and Social Networking*, *18*(9), 552–556.

Leinfelder, R. (2013). Assuming responsibility for the Anthropocene: Challenges and opportunities in education. *RCC Perspectives, 3*, 9–28.

Lewis, S. L., & Maslin, M. A. (2015). Defining the Anthropocene. *Nature, 519*(7542), 171.

Littman, J., & Wrubel, L. (2019). Climate change Tweets Ids. [Online] Available from: https://doi.org/10.7910/DVN/5QCCUU, Harvard Dataverse, V1.

Liu, F., Ford, D., Parnin, C., & Dabbish, L. (2017). Selfies as social movements: Influences on participation and perceived impact on stereotypes. *Proceedings of the ACM on Human-Computer Interaction, 1*(CSCW), 72.

McCain, J. L., Borg, Z. G., Rothenberg, A. H., Churillo, K. M., Weiler, P., & Campbell, W. K. (2016). Personality and selfies: Narcissism and the Dark Triad. *Computers in Human Behavior, 64*, 126–133.

McCain, J. L., & Campbell, W. K. (2018). Narcissism and social media use: A meta-analytic review. *Psychology of Popular Media Culture, 7*(3), 308–327.

McGill, M. (2018). How the light gets in: Notes on the female gaze and selfie culture. *Maifeminism* [Online] Available from: http://maifeminism.com/how-the-light-gets-in-notes-on-the-female-gaze-and-selfie-culture/ [Accessed December 2019].

Molloy, A. (2014). *'Selfie obsessed' teenager Danny Bowman suicidal after failing to. The Independent.* [Online] Available from: www.independent.co.uk/news/uk/home-news/selfie-obsession-made-teenager-danny-bowman-suicidal-9212421.html

Moody, E. J. (2001). Internet use and its relationship to loneliness. *CyberPsychology & Behavior, 4*(3), 393–401.

Moon, J. H., Lee, E., Lee, J. A., Choi, T. R., & Sung, Y. (2016). The role of narcissism in self-promotion on Instagram. *Personality and Individual Differences, 101*, 22–25.

Morozov, E. (2009). Iran: Downside to the "Twitter revolution". *Dissent, 56*(4), 10–14.

Murphy, K. (2015). What selfie sticks really tell us about ourselves. *The New York Times* [Online] Available from: www.nytimes.com/2015/08/09/sunday-review/what-selfie-sticks-really-tell-us-about-ourselves.html [accessed December 2019].

Nunes M. (2017). Selfies, self-witnessing and the 'out-of-place' digital citizen. In: A. Kuntsman (ed.) *Selfie Citizenship*, 109–117. Cham: Palgrave Macmillan.

O'Connor, A. (2014, August 1). *Emma Watson joins laughing protest against Turkish Deputy Prime Minister* [Online] Available from: www.dailyedge.ie/emma-watson-laughing-campaign-turkey-1600470-Aug2014/.

Oxford Dictionaries. (2013, November 18). Oxford Dictionaries word of the year 2013. [Oxford Words Blog]. [Online] Available from: http://blog.oxforddictionaries.com/2013/11/word-of-the-year-2013-winner/

Pattberg, P., & Zelli, F. (eds.). (2016). *Environmental politics and governance in the Anthropocene: Institutions and legitimacy in a complex world.* London: Routledge.

Pidcock, R., Pearce, R., & McSweeney, R. (2019). Mapped: How climate change affects extreme weather around the world. *Carbon Brief.* [Online] Available from: www.carbonbrief.org/mapped-how-climate-change-affects-extreme-weather-around-the-world [Accessed February 2020].

Pisani, L. (2015). Women and selfie culture: The selfie as a feminist communication tool. [Paper presentation]. Congress of the Humanities and Social Sciences, Ottawa, May 30 – June 5.

Pollock, G. (2003). *Vision and difference: Feminism, femininity and the histories of art.* London: Psychology Press.

Routh, P. (2016). The politics of transformation: Selfie production of the visually marginalised. In: A. Karatzogianni, D. Nguyen, & E. Serafinelli (eds.) *The digital transformation of the public sphere: Conflict, migration, crisis and culture in digital networks*. London: Palgrave Macmillan.

Russell, J., Cupp, L., & Macko, N. (2015). Selfies for self care: The role of "Narcissistic" media in the political action. [Unpublished manuscript] Scripps College, Claremont.

Rutledge, P. B. (2013). #Selfies: Narcissism or self-exploration. *Psychology Today*. [Online]. Available from: www.psychologytoday.com/gb/blog/positively-media/201304/selfies-narcissism-or-self-exploration [Accessed December 2019].

Ryan, E. G. (2013). Selfies aren't empowering. They're a cry for help. *Jezebel*. [Online] Available from: https://jezebel.com/selfies-arent-empowering-theyre-a-cry-for-help-1468965365 [Accessed December 2019].

Saatchi (2017). #SaatchiSelfie. *Saatchi* [Online] Available from: www.saatchigallery.com/selfie/ [Accessed December 2019].

Scranton, R. (2015). *Learning to die in the Anthropocene: Reflections on the end of a civilization*. San Francisco: City Lights Publishers.

Senft, T., & Baym, N. (2015). What does the selfie say? Investigating a global phenomenon, *International Journal of Communication*, *9*, 1588–1606.

Shane-Simpson, Christina, et al. (2020). "I love my selfie! An investigation of overt and covert narcissism to understand selfie-posting behaviors within three geographic communities." *Computers in Human Behavior*, *104*, 106158.

Shaw, S., & Lee, J. (2012). *Women's voices feminist visions* (2nd ed.). New York: McGraw-Hill.

Strauss, D. L. (2003). *Between the eyes: Essays on photography and politics*. New York: Aperture.

Sung, Y., Lee, J. A., Kim, E., & Choi, S. M. (2016). Why we post selfies: Understanding motivations for posting pictures of oneself. *Personality and Individual Differences*, *97*, 260–265.

Trudy. (2015). What's really going on with the sexist backlash against selfies. *Gradientlair* [online] Available from: www.gradientlair.com/post/115395306478/sexist-backlash-against-selfies.

Wang, J. L., Jackson, L. A., Wang, H. Z., & Gaskin, J. (2015). Predicting social networking site (SNS) use: Personality, attitudes, motivation and internet self-efficacy. *Personality and Individual Differences*, *80*, 119–124.

Weiser, E. B. (2015). # Me: Narcissism and its facets as predictors of selfie-posting frequency. *Personality and Individual Differences*, *86*, 477–481.

Wright, C., Nyberg, D., Rickards, L., & Freund, J. (2018). Organizing in the Anthropocene. *Organization*, *25*(4), 455–471.

10 Urbicide in the Anthropocene

Imagining Miami futures

Stephanie Wakefield

The idea that Earth has entered a new Anthropocene epoch is commonly linked to thesis of the planet's urbanization. With the world's majority now living in cities—a figure projected to increase to two thirds by 2050—along with being a time of climate crisis the human age is routinely characterized as the urban age (Kilcullen 2013; Brenner 2014; Brenner & Schmid 2014; Ruddick 2015). Here, capitalist globalization is understood to be turning Earth into a standardized, though uneven, interlinked network of social-spatial processes and technologies (Adams 2019). Productive of the majority of the world's greenhouse gas emissions and waste, urban systems are also understood as vulnerable to the Anthropocene's natural disasters, rising seas, and extreme heat (Swyngedouw 2015; Ritchie & Roser 2018–2019). But via a bricolage of resilience infrastructures, governments and planners hope that extreme events and threshold crossings will be absorbed and managed, allowing cities to maintain their basic structures and functions (Braun 2014; Wakefield & Braun 2014). In this way, it is imagined, the inexorable development of urbanization and the Anthropocene will remain necessarily companion processes. In fact for geologist Jan Zalasiewicz and colleagues (2014), cities will be one of humanity's longest lasting traces on the planet, future fossils whose imprint will remain long after humans. With the urban envisioned as the inevitable form of the twenty-first century, the only question mark it seems is whether cities will be more or less resilient or smart, inclusive or equitable.

However, concurrent with projections of planetary urbanization as a seemingly inevitable condition of the Anthropocene—or Urbicene, as Erik Swyngedouw (2017) has suggested it instead be named— different, more apocalyptic, visions of urban futures are emerging.[1] Along with the sheathing of Earth in an integrated tangle of urbanized life and infrastructure, the Anthropocene has also been marked by a wave of projections of the city's end. In these visions, the city's end comes not from the dissolving of discrete cities into an unbounded urban planet but as the result of natural disasters or human designs. Such imaginaries shift away from modern images of the

DOI: 10.4324/9781003128854-11

city and forward novel forms of the longstanding urban apocalypse trope. Perhaps, nowhere in the United States are such imaginaries more ubiquitous than Miami Beach, Florida.

Best known in the popular imagination as a subtropical urban paradise, Miami Beach continues to be tied to images of luxury, leisure, and pleasure, a geographical imaginary the city's real estate and government seek to maintain. But the recent onset of sea rise is transforming the city's imaginary. Built on porous limestone and one of the lowest-lying coastal metropolitan areas in the United States, Miami Beach is seen as ground zero for sea rise in the country. The Miami metropolitan region is already experiencing sunny day flooding saltwater coming up through the city's sewer system and sidewalks even on non-rainy days six times per year (Spanger-Siegfried et al. 2014; Wdowinski et al. 2016). Flooding is projected to increase to 80 times per year by 2030, and 380 times a year across a much expanded territory by 2045 (SFRCCC 2015; UCS 2016). Along with sea rise, hurricanes and extreme heat are also intensifying. As flooding has become more frequent in recent years, spectacle and speculative imaginaries of Miami Beach have shifted toward the apocalyptic, with media, government, academics, and planners increasingly envisioning the city's future destruction, due variously to natural forces or large-scale engineering projects.

In what follows, I trace these emergent Anthropocene images of Miami Beach. In doing so, I analyze their distinct but related depictions of the city's end, exploring the distinct forms of destruction imagined, types of human and nonhuman agency involved, and urban futures brought forward. First, the resilient city.

The resilient city

The City of Miami Beach's growth in the early twentieth century centered largely on speculative and spectacular imaginaries of a sunny urban fantasyland for Northeastern middle and upper classes (Nijman 2011, p. 23). Boosterist images in newspaper ads and tourism brochures of swaying palm trees and pristine beaches were essential to attracting real estate investment and development to the island. "To millions of Americans by the early 1920's," historian Gregory Bush (1999) recounts, Miami Beach was "associated with speeding cars and motorboats, spectacular stunts, grandiose hotels, mansions, lavish parties, polo matches, bathing casinos, and crowds of sunbathers" (p. 156). The spectacle of a glamorous urban paradise—which historically existed alongside Jim Crow racial segregation, antisemitism, and extreme economic inequality (Connolly 2014)—remains central to Miami Beach's reputation (Nijman 2011, p. 23).

But the recent onset of sea rise related-flooding and increasingly strong storms is transforming the city's geographical imaginary. Former Florida governor Florida Rick Scott publicly denied climate change—and issued an unofficial ban on use of phrases like "climate change," "global warming," and "sea level rise" in government documents (Korten 2015). However in recent years the compound threats climate change poses to South Florida's future have become governmental issues. The South Florida region is ranked first in the world in terms of assets vulnerable to sea rise, and the state stands to lose billions in property and tourism by 2030 due to climate change (Paustian-Underdahl et al. 2016, p. 26). Dependent on real estate and tourism, the City of Miami Beacha's resilience infrastructures aim to maintain urban property markets and investor confidence (Smiley et al. 2018; Wakefield 2019b; Grove et al. 2020). Governmental budgets in Florida are largely drawn from real estate tax. A major crash in property values in the area would reduce government funding, including that allocated for resilience projects (Grove et al. 2020). Smooth and successful action on sea rise is thus seen as key to protecting critical business and real estate assets and an increasingly important signal to banks, investors, and developers that the city remains financially viable.

Toward this end, in recent years academic and resilience planning networks have enrolled the city as a living laboratory for climate adaptation. Through scenario workshops and design charrettes, planners, scientists, and engineers seek to imagine the city's vulnerability to sea rise, flooding, and hurricanes, and to design modular infrastructures able to anticipate and mitigate these threats (Wakefield 2019b). In this vein, Miami Beach's $500 million resiliency infrastructure program, "Miami Beach Rising Above," includes a fleet of new industrial pumps, elevated streets, and seawalls to prepare the city for the next 30–50 years of sea rise, coastal flooding, and King Tides (ULI 2018; Wakefield 2019b). Likewise, the City of Miami, City of Miami Beach, and Miami-Dade County have each created Chief Resilience Officer positions to oversee municipal resilience planning. Their Greater Miami and the Beaches partnership recently released Resilient305, a resiliency strategy plan for fielding a range of increasingly frequent shocks and stressors, from sea rise and hurricanes to socioeconomic inequality and traffic congestion (Grove et al. 2020).[2] In addition to large-scale municipal efforts, programs such as the South Florida Disaster Resilience Initiative are working on programs to increase the capacities of the city's poor and vulnerable communities to prepare for and recover from extreme storm events.

Imagined across these resilience initiatives is a future city that, although increasingly inundated by seas and storms, remains both viable and governable. The water will come, but through adaptive infrastructures, its impact will be managed. Existing urban infrastructure and terrain will be transformed, retrofitted, or supplemented, but broader system parameters will be maintained amidst turbulence. Circulation of capital, tourism, and investment will be secured, by buffering the catastrophic effects these structures create. Through a bricolage of modular, integrated green and gray infrastructures able to absorb and field and even gain from extreme events (Rodin 2015), the global city of trade, tourism, and finance will ward off and even thrive while seas rise up through its foundations and waves crash against its shores.

The "deus ex machina" of the Anthropocene, as geographer Kate Derickson (2018, p. 426) puts it, the resilient city is thus a city that, by both allowing and warding off ecological apocalypse, can still be saved. Large-scale ecological and technical infrastructural transformations work together with situated local adaptive capacities of the poor and vulnerable, together constituting an eco-cybernetic, adaptive urban terrain (Wakefield & Braun 2014; Wakefield 2020). This revised spatial-governmental form is seen as promising a future urban immunized from extreme press and pulse events, both ecological and social (Adams 2014). Rather than twentieth-century visions of stability or progress, the resilient city will be a "crisis city" (Gotham & Greenberg 2015), perpetually vulnerable to environmental compound risks. But it is still a governable, savable city. With crisis as a constitutive backdrop, the resilient city's adaptive infrastructures enable it to remain a financially viable global city, subtended by and interlinked with the critical flows of capital, people, commodities that constitute neoliberal planetary urbanization. In the resilient urban, what is seen as still possible is the continued reproduction of the city's basic social, economic, and geographical parameters amidst ecological catastrophe. This future vision is aptly summed up in the name of Miami's $400 million resiliency infrastructure bond: "Miami Forever."

Goodbye, Miami

A second, more apocalyptic imaginary of Miami's future has become commonplace in recent years however. Rather than imagining human ability to control catastrophe, as does resilience, this imaginary centers on the city's imminent and inevitable demise at the hands of natural forces. Forwarded by an increasing number of journalists and artists, this vision most often involves visions of natural forces overwhelming human agency. Originally titled "Goodbye, Miami," journalist Jeff Goodell's 2013 *Rolling Stone* article set the tone for a series of apocalyptic Miami media pieces to come. Along with the late geologist Peter Harlem's 2008 mapping of Florida's future sea rise flooding and Al Gore's 2010 *An Inconvenient Sequel* Miami Beach feature, Goodell's article helped lay the visual basis for Miami's problematization (Foucault 1984) as the frontline of sea rise in the United States. In the essay, Goodell imagined the city's demise in 2030 at the hands of a fictional hurricane. The city's famous palm trees are obliterated by wind. Storm surges of 24 feet annihilate Ocean Drive's historic art deco hotels. White sand beaches are left littered with trash and human waste. Celebrity mansions are flooded. Salt water corrodes electrical infrastructure, leaving the city in darkness for months.

Similar visions of a city and its human residents overwhelmed by nature's powerful forces frequently populate climate change-themed art produced in the Miami area: people floating in tanks of water (Jan 2015), lawn placards indicating the home's future sea rise risk (Cortada 2018), a sand-covered traffic jam installed along Ocean Drive (Erlich 2019). Forwarded in this imaginary is a vision of the city overwhelmed by powerful natural forces. Human efforts are revealed to be too little, too late. These images reverse modern visions of humans as powerful agents and the natural world as object to control and order. Instead, it is nature that comes to decimate humans and the human hubris that is the city. Rather than powerful agents, in Goodell's account humans are portrayed as collateral damage (storm casualties) and bumbling, science-denying, bible-quoting fools. Critical infrastructural hopes of maintaining business-as-usual control lie in ruins. Nature is the real force.

Imagining the modern city's end has long fascinated artists and writers, for whom envisioning the obliteration of skyscrapers and landmark attractions of places like New York has enthralled, entertained, or offered a compelling means of imagining the overturning of oppressive social relations that only apocalypse seems able to achieve (Page 2010). In contrast, Goodbye, Miami's vision of the end is pedagogical, intended to serve as a warning and call to action for policymakers and planners. In this vein, following his apocalyptic vignette Goodell includes a brief lesson on global warming and sea rise and their projected impacts over the coming century. But Goodbye, Miami also aims to reveal and correct the modern human/nature binary seen as having created climate change. The city's imminent demise here is the result not only of swelling seas and crushing waves but of hubristic modern idea of nature. "Even more than Silicon Valley," Goodell writes,

> Miami embodies the central technological myth of our time – that nature can not only be tamed but made irrelevant. Miami was a mosquito-and-crocodile-filled swampland for thousands of years, virtually uninhabited until the late 1800s. Then developers arrived, canals were dug, swamps were drained, and a city emerged that was unlike any other place on the planet, an edge-of-the-world, air-conditioned dreamland of sunshine and beaches and drugs and money...Floods would come and go and hurricanes might blow through, but the city would survive, if only because no one could imagine a force more powerful than human ingenuity. That defiance of nature – the sense that the rules don't apply here – gave the city its great energy. But it is also what will cause its demise.
>
> (n.p.)

In many ways Goodbye, Miami resonates with recent critical work around nonhuman natures and vital materialism (Braun 2008). Seeking to correct anthropocentric modern thinking—the hubristic ethos also embodied in resilience infrastructure— scholars in this vein demonstrate how, rather than an inert object, the nonhuman world is agential. Donna Haraway (1988), for example, advocates acknowledging objects as active entities as part of overturning capitalist patriarchy and its view of nonhumans as resources. More recently, Jane Bennett (2007) has sought to disrupt the modernist partitioning of the world into inert nonhumans and lively humans by lavishing attention on the "thing-powers" of worms or trash, which she argues constitute vibrant political actors in their own right, part of human collective publics akin to what Bruno Latour (1993) similarly calls a "parliament of things." Likewise Anna Tsing (2015) argues that modern Enlightenment views of nature as a passive backdrop have blocked attention to lessons from entangled, lively natures such as the matsutake mushroom whose "riotous colors and smells" (p. 1) thrive in disturbed landscapes.

While equally pedagogical, Goodbye, Miami also pushes further beyond these attempts to overcome the modern human/nature binary. For thinkers like Tsing, Haraway, and Latour, undoing modern Enlightenment approaches means understanding that humans and nonhumans are constitutively intertwined and developing collaborative relations with them. However for other scholars such as Nigel Clark (Clark 2010; Clark, Stasch, & Bialecki 2020), while humans and nature may be far more interrelated than modern approaches often allow, this relationship is not one of equality or simple co-constitution. Rather, for Clark, natural forces far exceed human aims and often do so in violent and destructive ways (Wakefield, Grove, & Chandler 2020). "Geophysical events – with their largely inhuman forces and timeframes – grate against assumptions of mutuality and co- presence," writes Clark (Clark, Stasch, & Bialecki 2020, p. 9).

Similarly, far from a collaborating entity in a "parliament of things," in Goodbye, Miami the urban ground becomes a hostile force. Sands mined in Central Florida and trucked in to stock Miami's famous beaches bury hotel lobbies. The glittering blue sea, normally pacified and flat, bulldozes historic art deco architectures into rubble. The city's limestone substrate, rather than invisible ground underfoot, permits salt water to infiltrate up through city streets and drains, rendering flood barriers' technologies used in other coastal cities ineffective. Nature becomes a military force and moral arbitrator. Writes Goodell (2013), describing one of the region's characteristic tropical afternoon storms, "it feels biblical, as if God were punishing the good citizens of Miami Beach for spending too much time on the dance floor" (n.p.).

From this vantage point, resilience efforts like those discussed in the preceding section are ill-fated attempts to prolong this human hubris (Neyrat 2019; Wakefield, Grove, & Chandler 2020). And they will ultimately, Goodell concludes, fail:

> Rising seas will present an escalating series of challenges, most of which, on their own, will appear to be manageable. It's not hard to see how it will play out: As each new crisis arises, engineers will propose expensive solutions and people may be fooled into thinking that sea-level rise is not such a big deal. But in many cases, sea-wall extensions and elaborate

> pumping and drainage systems will turn out to be giant boondoggles, with money shoveled out to politically connected contractors for projects that are ineffective or overwhelmed by continually rising seas. (n.p.)

In the long term, Goodell concludes, the twentieth-century engineering hubris on which the city was built—draining swamps to build suburbs, attempting to control regional water cycles, building coastal nuclear plants exposed to rising seas and hurricanes—will be its downfall. Conventional sea walls and flood barriers will be useless, Goodell envisions. Well before seas or deadly storms submerge real estate, salt water will infiltrate the region's drinking water supply via Biscayne Aquifer, making the city unlivable. Instead of washing dirty money or housing international elites, high rises built during the city's post-Great Recession building boom are abandoned to drug dealers and crocodiles. In the longer term, investors leave and tourism plummets. Roads become undrivable at high tides. Banks stop issuing mortgages. By 2100, "rising sea levels will turn the nation's urban fantasyland," Goodell concludes, "into an American Atlantis...a popular snorkeling spot where people could swim with sharks and sea turtles and explore the wreckage of a great American city" (n.p.).

The Islands of South Florida

A third vision of the city's future, however, is coming into view. Many Miami planners and politicians agree with Goodell that current resiliency efforts are insufficient, in the long term, to manage projected sea rise. According to Miami Beach's former city engineer Bruce Mowry, who oversaw many of the resilience infrastructure projects discussed above, whether or not carbon emissions are reduced today, coastal areas remain vulnerable. The City's $500 million resiliency infrastructures are only designed to secure the city for 30–40 years. After that, as University of Miami geologist Harold Wanless sees it, the city is "doomed."

These increasingly common views are in large part based on the challenges of the Miami metropolitan region's geology. Built on porous limestone, made of marine deposits accumulated from the remains of sea creatures over millions of years when the area was underwater, the city faces sea rise flooding not only along its coasts but also from water coming up through its foundation. During the city's annual King Tides, periods of especially high tides, sea water comes up through street drains and streets and floods into roadways and intersections. Thus, whereas cities in the Netherlands use dykes to mitigate rising seas or Manhattan might be salvaged using massive Bjarke Ingels-designed sea barriers, Miami's permeable foundation means flooding will advance with or without walls. As summed up by former Speaker of the Florida House of Representatives and director of the

Institute of Environment at Florida International University Tom Gustafson (2018):

> Given current South Florida elevations, GHGs already in the atmosphere, the likely acceleration of such GHG emissions for the next twenty years and more, and expected sea level rise feedback mechanisms, there is little likelihood that South Florida, as we know it today, will be habitable by the year 2100 AD unless we are willing to start very soon to build something that creates communities elevated about current conditions. (n.p.)

Rather than forefront the city's destruction as in Goodbye, Miami, many politicians, engineers, and scientists now advocate laying the ground for a different, more aquatic strategy for adapting to future environmental change. In public and private conversations around the city, a consensus is emerging that human life can continue in the Miami area only if the city is radically transformed. Toward this end, a proposal is circulating among politicians, engineers, and scientists for physically transforming the city into an infrastructurally linked, self-sufficient zone of high grounds, connected like the Florida Keys by new elevated bridges (AST et al. 2017). As envisioned by Gustafson, the design's most vocal promoter, the "Islands of South Florida" will be constructed by converting 100% of the city's currently developed land into fill to create a series of new elevated islands—the Islands of South Florida (Crowl & Teutonico 2018). On the islands will be luxury high-rise buildings, with space allocated for a service sector workforce. The rest of the current South Florida population will be migrated out via forms of planned retreat. The artificial islands, in Gustafson's vision, will be based in localized food and materials production and water-based lifestyles.

In this vision, we can see the emerging glimpses of a potential new paradigm of Anthropocene coastal adaptation that, while related, must also be distinguished from resilience. The resilience visions described in the first section of this chapter work on and remodulate urban coasts and built environments to secure existing social, economic, and geographical conditions. While urban resilience departs in significant ways from modern engineering-based planning, imagining instead a bricolage of situated social, technical, and ecological infrastructures, it is still fundamentally modern because it maintains hope that the city can be saved (Chandler 2019). Through an interlaced, layered suite of soft and hard infrastructures to field and absorb disaster events—mangrove buffers, elevated roads, pumps, and sea walls—existing urban geographies remain viable and governable.

The South Florida Islands, however, suggests the impossibility of this resiliency vision. In this emergent form of coastal adaptation—what we might call "stage 2 adaptation"—the existing urban formation is viewed as doomed. Accordingly, rather than retrofitted or adapted, the city itself must be preemptively destroyed. Envisioned here is not the elevation of Miami Beach—as engineer Mowry puts it, "defend[ing] against the water at the present interface of the ocean to the land" (AST et al. 2017)—the raising of roads and eventually a whole city above sea level as in the city's resilience efforts. Instead, SFI envisions the necessity of pulverizing the current city in order to use its substrate as fill for an entirely new geography.

This approach can be understood as a kind of anticipatory Anthropocene urbicide. The military tactic of deliberately destroying ("killing") entire cities, urbicide is a term that has been used to describe American military's walling off towns and urban districts in Iraq or the Israeli Defense Force's bulldozing of Jenin camps (Weizman 2007; Graham 2011). Frequently associated with counterinsurgency, urbicide is often described as a strategy of attacking an insurgent force's "urban foundations" (Graham 2011, p. 226). Critical infrastructure, housing blocks, governmental facilities, and cultural institutions are seen by attacking forces such as the IDF as the ground upon which insurgencies, like the Palestinian resistance, grow and flourish.

In the South Florida Islands, however, the physical ground appears as a more pragmatic and existential problem as well as solution: the raw material for a new territory, itself recently engineered. Miami was built on drained swampland. Likewise, the current topography of Miami Beach, a thin barrier island in the late nineteenth century, was created in the early twentieth century as part of a real estate scheme by entrepreneur Carl Fisher. Under Fisher—who refused to sell the Miami Beach property he would build to Jews (Carson 1955)—workers, elephants, and mules cleared the island's dense mangrove swamps with axes and saws. Millions of cubic yards of fill made from sand dredged from adjacent Biscayne Bay and Everglades soil was added atop the existing limestone foundation (Mohr 2015). Layers of sand and limestone were arranged into a relatively level terrain approximately 3–5 feet above sea level. On them the streets and sidewalks of Miami Beach were built with rock brought in by barge from inland mines, providing the grounds for Fisher's racially restricted elite crowd of "newly rich industrial magnates" (Carson 1955).

Carrying this approach forward into the Anthropocene, in Gustafson's view, government must obtain 100% of the city's developed land to use as fill. This will be done via multiple methods. Some land must be condemned. A smaller percentage, he argues, can be sold to private developers and subsequently bought out as well. As is done in the limestone mining operations that dot Florida, conceivably this would involve clearing the earthen and built materials that would not be used. Iconic palm trees would be replanted or mulched. Historic art deco hotels would be trucked out, preserved elsewhere, or demolished. Once cleared, excavators would remove the soft sandy top layer. Explosive dynamite would break up hard areas of limestone. Both would be crushed into a uniform size suitable for use as fill.

Like many instances of urbicide, imagined in South Florida Islands is the demolition of the existing urban form—its infrastructure and geological ground—and replacement with a form chosen by the offensive force. In Iraqi cities, for example, the aim of American military sought to install a liberal, American regime of life and control. And as Graham (2011) notes, urbicide is not only linked to wartime but also peacetime efforts by neoliberal elites, especially in the Global South, to rebrand as "global cities" by bulldozing slums, evicting dwellers, and building in their place financially attractive highways, luxury high rises, and stadiums (p. 84). In a similar but more existential-pragmatic vein, South Florida Islands envisions converting urban ground into a fortified, aqua-urban island territory.

Importantly, this envisioned geographical form is not a continuation of planetary urbanization, albeit elevated. Nor even is it a continuation of what architect Ross Exo Adams (2019) names urbanization, the "eclipsing" of the urban via the attempted project of interlinking of circulatory flows of people, commodities, and information into a single, continuous system of social, technological, and economic control. What is instead envisioned is the destruction of existing urban form to create a new geography based on *delinking* from such global networks. Here what Graham (2005) has described as "switching cities off"—military targeting of critical urban infrastructures—takes on new meaning. For Gustafson, the Islands of South Florida must redesign and localize all of the infrastructures that previously subtended Miami life. He explained in an interview conducted during the beginning weeks of the COVID-19 pandemic:

> We have to cut all global supply chains. You see Coronavirus? The future requires isolationism.
>
> (personal communication 2020)

Here the island appears as a key figure of the Anthropocene, albeit not as a symbol of relationality (Chandler & Pugh 2018; Pugh 2018) but of detachment. Among other infrastructural transformations, in Gustafson's (2018) design the islands must "grow locally most or all our food; manufacture locally most or all our goods and building materials in daily use; and, secure the raw materials/durable goods we might need" (n.p.). Production would further be localized to be something residents participate in themselves: "we need tabletop design and manufacturing tools," he explains (personal communication 2020). Buffered by mangrove trees and other salt water-resistant vegetation, the islands would feature extreme storm-resistant architecture able to withstand the hammering rain, storm surge, and powerful winds expected in the region. Power will be generated through wind and solar technologies supplanted by fourth-generation nuclear. The Islands will maintain an electronic "Library of Alexander" and seed bank of crops suited to the region's long-term growing conditions. What these infrastructural relocalizations suggest is not the immunization of existing forms of urbanization, but the creation of a new geographical form in opposition to urbanization itself.

Conclusion

The end of the city is not new. For geographer Sarah Knuth (2020), this trope is endemic to modern urban regimes, a fact that can be seen in the myriad targeted interventions into urban form and longevity. Urban austerity, de-industrialization, and ghettoization, like gentrification or touristification, target and destroy urban geographies (Sugrue 2005; Gotham & Greenberg 2015). From creative destruction to "ruination-as-usual," Knuth describes capitalist urban regimes' propensity for devaluation and ruination in terms of Marx's classic formulation: "all that is solid melts into air." Along with concrete destruction of urban forms, "eschatological evocation[s] of urban apocalypse" (Gandy 2005) are as old as the city itself, and feature prominently in future visions of Anthropocene urbanization. Megacities like Lagos, for example, are frequently characterized as hyperbolically as hotbeds of "poverty, violence, disease, political corruption, uncontrollable growth and manic religiosity" (Gandy 2005), images of crowded, chaotic, urban twenty-first century to come (Kilcullen 2013). Such apocalyptic descriptions likewise have a history in Miami. In the late 1970s and early 1980s, the city was labeled "Murder City, U.S.A., where crime had gone 'beserk'" and cocaine trafficking reigned (Nijman 2011, p. 63). Writes historian Bush (1999),

> The regional low point probably came after the McDuffie riots and the Mariel boatlift in 1980, when Times cover article featured the area as "Paradise Lost." Had the promotional bubble burst with the exposure of racial brutality in the police department, broken promises to the African American problems that had been so long overlooked by the white power structure? Hardly. There was always another hastily devised "Action Plan" to thwart grievances and promote the area amid uncertain economic times.
>
> (p. 170)

But the Anthropocene is also producing distinct visions of the city's end, apocalyptic imaginaries that are novel both for Miami and cities more generally. As I have traced in this chapter, these urbicidal visions foresee the city's demise in diverse ways. Urban resilience designs are currently the city's most widely accepted approach, one embraced ubiquitously by governments and designers. This resilience approach seeks to manage sea rise and flooding via modular adaptations and technological remodulations to produce a hoped-for "Miami Forever," part of a more expansive process of resilient urbanization. In contrast, Goodbye, Miami, a popular media imaginary, portrays these eco-cybernetic resilience aspirations as a doomed repetition of the modern Promethean attempt to control nature. Eventually, this imaginary envisions, nature will return to destroy the city. In this nature-based urbicide, the destructive power of natural forces reduces human aims and structures to rubble. Left in place of the urban is a ruin, slowly overrun by nature. In the third imaginary, the South Florida Islands, maintaining life upon existing geo-urban boundaries and foundations—whether through sea walls or through soft ecological infrastructure, as in the currently dominant urban resilience vision—is declared impossible. Here we saw a second form of Anthropocene urbicide. Rather than abdicate the city to nature however, here it is politicians and engineers who advocate the destruction of Miami's current geography, in order to repurpose it as fill for a self-sufficient elevated island territory.

While Goodbye, Miami painted a somewhat standard Anthropocene image—urban structures submerged underwater—South Florida Islands, I believe, sketches out something more novel: the preliminary outlines of a paradigm of coastal adaptation beyond urban resilience. Rather than secure existing land formations and urban boundaries against sea rise (as resilience seeks to do), here we saw the preemptive killing of an iconic but doomed modern urban form, and the creation of a new island territory: locally self-sufficient in terms of producing basic needs, but interconnected regionally in terms of security and infrastructure. In South Florida Islands, all that which is solid—not just buildings, but Miami's existing urban form itself—does not melt into air. Instead, urbicidally obtained, the city's physical ground becomes the raw input for creating a novel geographical form.

These new imaginaries of Miami are important in their own right, remodulating the city's geographic imaginary as well as producing physical infrastructural reconfigurations. And as I have argued, they suggest the emerging glimpse of a new paradigm of coastal adaptation beyond urban resilience. But these visions also raise questions about the supposed urban fate of the Anthropocene, and the proclaimed inevitability of a certain form of urban life in the twenty-first century. Most obviously, they suggest the possibility that the "'lifetime' of today's urban fabrics and metropolitan forms," as Knuth puts it, may be more ephemeral than many accounts of the Anthropocene imagine. The changing environments and adaptive challenges of the twenty-first century may well lead to the destruction of seemingly unquestionable spatial forms such as the urban, and the creation of new, previously unimagined geographies. Such an urbicidal Anthropocene might encompass the brutal destructive agency of nonhuman forces, in ways that would not be comprehensible in terms of relational symmetry or entanglement or governmental enrollment. Likewise, Anthropocene urbicide would also suggest the emergence of powerful forms of human agency directed toward shedding modern forms as fundamental as the urban itself and using them as raw materials for new spatial forms. Such creations would clearly be multiple in form. But South Florida Islands suggests that at least one potential end of the city might be tied to a process of delinking, rather than (or concurrent with) the further eco-cybernetic interlinking suggested by the planetary urbanization thesis.

Whether visions such as South Florida Islands, if accomplished, would lead to more desirable ways of living or offer a means to break with existing infrastructural forms of control remains to be seen. The fact that SFI breaks formally with aspects of the interconnected global system of infrastructural control that Adams (2019) calls urbanization does not on its own give one reason to believe it would be a project of leaving the liberal capitalist system of control. In fact in Gustafson's vision SFI will remain tied in other ways to neoliberal governance. As imagined by Gustafson, the islands will also act as a frontline US military bulwark against the "pirates of the Caribbean" which he sees as arising in the future with the failure of some island states. For this some islands will be home to military bases, connected by massive new bridges built for smart electric vehicles and automated trucks to move supplies and people throughout the islands. Perhaps instead of leaving neoliberal regimes behind, the urbicidal Anthropocene and attendant territorialization instead offers a glimpse of potential recalibrations to current governmental formations that will emerge as liberal regimes seek to maintain themselves amidst the shocks and stresses of the coming century. Far from the utopian floating cities imagined by the likes of the Seasteading Institute, this resonates with potentially darker visions of a future made up of relocalized bunker zones for the wealthy elite and their workers, fortified against environmental and human surrounds.

These ends of the city are at present just hypothetical thought experiments. Nevertheless, allowing them to momentarily suspend the sense of inevitability that surrounds urbanization in the Anthropocene opens thought to the possibility that cities are not the only possible future. In this opening, many questions become important. What spatial forms may come to replace the urban? What new modes of government and forms of life might such geographies produce? Is it possible to proactively imagine desirable new spatial forms beyond the urban, rather than leaving them to always emerge reactively as environmental adaptations or as top-down governmental interventions? Exploring these questions opens important new pathways of analysis for critical urban scholars and will require new conceptual tools appropriate to changing environments.

Notes

1 Franciszek Chwałczyk (2020) has similarly suggested the term "Urbanocene."
2 www.mbrisingabove.com/wp-content/uploads/Resilient305_final.pdf.

References

Adams, R. E. (2014) Natura urbans, natura urbanata: Ecological urbanism, circulation, and the immunization of nature. *Environment and Planning D: Society and Space*, vol. 32, no. 1, pp. 12–29.

Adams, R. E. (2019) *Urbanization and circulation*, London: Sage.

Alliance of the Southern Triangle (AST), Mowry, B., Obeysekera, J., Robinson, K.S., Stoddard, P. and Zuluaga, N. (AST et al.) (2017) Future cities: Miami research intensive public session. *Miami Rail* 23. https://miamirail.org/issue-23/future-cities-miami-research-intensive-public-session/.

Bennett, J. (2007) *Vibrant matter: A political ecology of things*, Durham, NC: Duke University Press.

Braun, B. (2008) Environmental issues: Inventive life. *Progress in Human Geography*, vol. 32, no. 5, pp. 667–679.

Braun, B. (2014) A new urban dispositif? Governing life in an age of climate change. *Environment and Planning D: Society and Space*, vol. 32, no. 1, pp. 49–64.

Brenner, N. (ed.) (2014) *Implosions/explosions: Towards a study of planetary urbanization*, Berlin: Jovis.

Brenner, N. and Schmid, C. (2014) The "urban age" in question. *International Journal of Urban and Regional Research*, vol. 38, no. 3, pp. 731–755.

Bush, G. (1999) "Playground of the USA": Miami and the promotion of spectacle. *Pacific Historical Review*, vol. 68, no. 2, pp. 153–172.

Carson, R. L. (1955) Forty years of Miami Beach, Tequesta. Miami, Florida. *Journal of the Historical Association of Southern Florida*, vol. XV, pp. 3–27.

Chandler, D. (2019) The death of hope? Affirmation in the Anthropocene. *Globalizations*, vol. 16, no. 5, pp. 695–706.

Chandler, D. and Pugh, J. (2018) Islands of relationality and resilience: The shifting stakes of the Anthropocene. *Area*, vol. 52, pp. 65–72.

Chwałczyk, F. (2020) Around the Anthropocene in eighty names—Considering the Urbanocene proposition. *Sustainability*, vol. 12, no. 4458, pp. 1–33.

Clark, N. (2010) *Inhuman nature: Sociable life on a dynamic planet*, London: Sage.

Clark, N. (2020) Can we have our nature/culture dichotomy back please? In Rapport, N. and Wardle, H. (eds.) *An anthropology of the enlightenment: Moral social relations then and today*, London: Bloomsbury, pp. 99–117.

Connolly, N. D. B. (2014) *A world more concrete: Real estate and the remaking of Jim Crow South Florida*. Chicago: University of Chicago Press.

Cortada, X. (2018) *Underwater HOA* [Public art exhibition] Village of Pinecrest, Miami, FL.

Crowl, T. and Teutonico, R. (2018) As the sea rises, South Floridians will get thirsty before they get wet. *The Invading Sea*, www.theinvadingsea.com/2018/06/04/as-the-sea-rises-south-floridians-will-get-thirsty-before-they-get-wet/

Derickson, K. D. (2018) Urban geography III: Anthropocene urbanism. *Progress in Human Geography*, vol. 42, no. 3, pp. 425–435.

Erlich, L. (2019) [Public art installation] City of Miami Beach, Miami Beach, FL.

Foucault, M. (1984) Polemics, politics, problematizations: An interview with Michel Foucault. In Rabinow, P. (ed.) 1998 *Essential works of Foucault, Volume 1*, New York: The New Press, pp. 111–119.

Gandy, M. (2005) Learning from Lagos. *New Left Review*, vol. 33, May–June, pp. 36–52.

Goodell, J. (2013) Miami: How rising sea levels endanger South Florida. *Rolling Stone*, www.rollingstone.com/feature/miami-how-rising-sea-levels-endanger-south-florida-200956/

Gotham, K. and Greenberg, M. (2015) *Crisis cities: Disaster and redevelopment in New York and New Orleans*, Oxford: Oxford University Press.

Graham, S. (2005) Switching Cities off. *City: Analysis of urban change, theory, action*, vol. 9, no. 2, pp. 169–194.

Graham, S. (2011) *Cities under siege: The new military urbanism*, London: Verso.

Grove, K., Cox, S., and Barnett, A. (2020) Racializing resilience: Assemblage, critique, and contested futures in greater Miami resilience planning. *Annals of the American Association of Geographers*, vol. 110, no. 5, pp. 1613–1630.

Gustafson, T. (2018) Islands of South Florida, abstract. *University of Florida Water Institute Symposium*, http://archives.waterinstitute.ufl.edu/symposium2018/downloads/Presentations/Gustafson.pdf

Haraway, D. (1988) Situated knowledges: The science question in feminism and the privilege of partial perspective. *Feminist Studies*, vol. 14, no. 3, pp. 575–599.

Harlem, P. (2008) Sea level rise map projections. *Sea Level Rise Collection*, 2, https://digitalcommons.fiu.edu/sea_level_rise/2

Jan, L. (2015) *Holocenes* [Performance art exhibition] Miami Dade College, Miami, FL

Kilcullen, D. (2013) *Out of the mountains: The coming age of the urban guerilla*, Oxford: Oxford University Press.

Knuth, S. (2020) All that is solid…. *City*, vol. 24, no. 1–2, pp. 65–75.

Korten, T. (2015) In Florida, officials ban term "climate change," *Miami Herald*, March 8, www.miamiherald.com/news/state/florida/article12983720.html

Latour, B. (1993) *We have never been modern*, Cambridge, MA: Harvard University Press.

Mohr, T. (2015) Dredge. In Elkin, R. (ed.) *South Florida rise and sink: The case of Miami beach*, Cambridge, MA: Harvard GSD Office of Urbanization.

Neyrat, F. (2019) *The unconstructable earth*, Bronx: Fordham University Press.

Nijman, J. (2011) *Miami: Mistress of the Americas*, Philadelphia: University of Pennsylvania Press.

Page, M. (2010) *The city's end: Two centuries of fantasies, fears, and premonitions of New York's destruction*. New Haven, CT: Yale University Press.

Paustian-Underdahl, S., Parra, C., Mesia, R., Papadopoulos, A., Scanlon, N., and Troxler, T. (2016) Getting ahead of the curve for a resilient economy: An introduction to risks and opportunities of climate change for top business sectors in Southeast Florida. *Sea Level Solutions Center* Publication, no. 8.

Pugh, J. (2018) Relationality and island studies in the Anthropocene. *Island Studies Journal*, vol. 13, no. 2, pp. 93–110.

Ritchie, H. and Roser, M. (2018–2019) *Our world in data: Urbanization*, https://ourworldindata.org/urbanization

Ruddick, S. (2015) Situating the Anthropocene: planetary urbanization and the anthropological machine. *Urban Geography*, vol. 36, no. 8, pp. 1113–1130.

Smiley, D., Harris, A., and Flechas, J. (2018) Levine is campaigning on climate change. But hissea-rise legacy is complicated. *Miami Herald*, May 24, www.miamiherald.com/article211326029.html

Southeast Florida Regional Climate Change Compact (SFRCCC) (2015) *Unified sea level rise projection for Southeast Florida* www.southeastfloridaclimatecompact.org/wp-content/uploads/2015/10/2015-Compact- Unified-Sea-LevelRise-Projection.pdf

Spanger-Siegfried, E., Fitzpatrick, M. F., and Dahl, K. (2014) *Encroaching tides: How sea level rise and tidal flooding threaten US East and Gulf Coast communities over the next 30 years*. Cambridge, MA: Union of Concerned Scientists.

Sugrue, T. (2005) *The origins of the urban crisis: Race and inequality in postwar Detroit*. Princeton, NJ: Princeton University Press.

Swyngedouw, E. (2015) Urbanization and environmental futures: Politicizing urban political ecologies. In Perreault, T., Bridge, G., and McCarthy, J. (eds.) *The Routledge handbook of political ecology*, London: Routledge, pp. 609–619.

Swyngedouw, E. (2017) More-than-human constellations as immuno-biopolitical fantasy in the Urbicene. In Gomez-Luque, M. and Jafari, G. (eds.) *New geographies 09: 'Posthuman'*, Graduate School of Design, New York: Actar Publishers, pp. 18–23.

Tsing, A. (2015) *The mushroom at the end of the world: On the possibility of life in capitalist ruins*, Princeton, NJ: Princeton University Press.

Union of Concerned Scientists [UCS] (2016) *Encroaching tides in Miami-Dade County, Florida.* Fact sheet: Sea level rise and tidal Flooding along the East Coast, www.ucsusa.org/sites/default/files/attach/2016/04/miami-dade-sea-level-rise-tidal-flooding-fact-sheet.pdf

Urban Land Institute (2018) Stormwater management and climate adaptation review. *An Urban Land Institute Advisory Services Panel Report*, Miami Beach, FL. April 16–19, www.mbrisingabove.com/wp-content/uploads/2018/04/Miami-Beach_ Panel-Report_lo-res.pdf

Wakefield, S. (2019a) Making nature into infrastructure: The construction of oysters as a risk management solution in New York City. *Environment and Planning E: Nature and Space*, vol. 3, no. 3, pp. 761–785.

Wakefield, S. (2019b) Miami Beach forever? Urbanism in the back loop. *Geoforum*, vol. 107, pp. 34–44.

Wakefield, S. (2020) Urban resilience as critique: Problematizing infrastructure in post-Sandy New York City. *Political Geography*, vol. 79, pp. 102148.

Wakefield, S. and Braun, B. (2014) Governing the resilient city. *Environment and Planning D: Society and Space*, vol. 32, no. 1, pp. 4–11.

Wakefield, S., Grove, K., and Chandler, D. (2020) The power of life. In Chandler, D., Grove, K., and Wakefield, S. (eds.) *Resilience in the Anthropocene: Governance and politics at the end of the world*, Abingdon: Routledge, pp. 1–22.

Wdowinski, S., Bray, R., Kirtman, B. P., and Wu, Z. (2016) Increasing flooding hazard in coastal communities due to rising sea level: Case study of Miami Beach, Florida. *Ocean and Coastal Management*, vol. 126, pp. 1–8.

Weizman, E. (2007) *Hollow land: Israel's architecture of occupation*, London and New York: Verso.

Zalasiewicz, J., Williams, M., Waters, C., Barnosky, A., and Haff, P. (2014) The Technofossil Record of Humans. *The Anthropocene Review*, vol. 1, no. 1, pp. 34–43.

11 Triggering the apparitions

Spectres of chemical seascapes

María Soledad Castro Vargas and Diana Barquero Pérez

Chemical geographies in the Anthropocene

Synthetic chemicals are intrinsic to the Anthropocene's landscapes (Figure 11.1). Industrialization represented a historical landmark, as it devised the fabrication and dissemination of synthetic chemicals capable of disrupting living organisms. Since World War II, their manufacture expanded geographically and increased exponentially, pointing to a dramatic phase of the Anthropocene (Arcuri and Hendlin, 2019). The rising production of these substances is not only considered one of the main agents of global change, but also evidences a clear imprint of the Anthropocene (Bernhardt, Rosi and Gessner, 2017).

Figure 11.1 Altered photograph from Carlos Flores's personal photo archive, 2018.

DOI: 10.4324/9781003128854-12

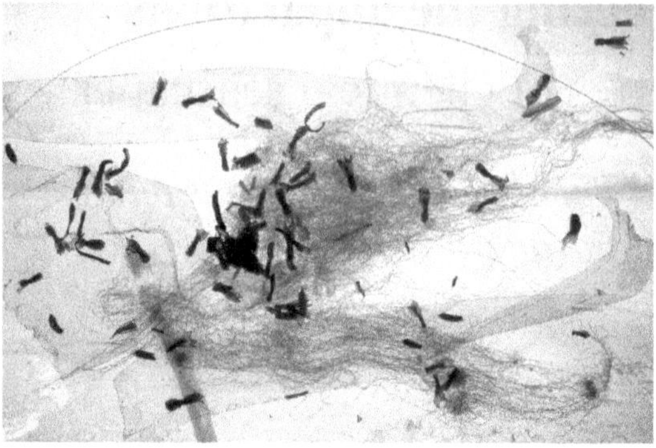

Figure 11.2 Altered photograph from Diana Barquero's personal photo archive, 2018.

During the past decades, contaminants of emerging concern (CECs) appeared in the scenario. CECs represent a novel environmental and health risk due to their adverse impacts (Naidu, 2016). Pharmaceuticals (e.g. Ibuprofen, tranquilizers, antidepressants), lifestyle compounds (e.g. caffeine, sweeteners, nicotine), personal care products (e.g. UV filters present in sunscreens and makeup, fragrances), drugs (e.g. cocaine, heroin), food additives and hormones, among others are classified as CECs. These compounds are present in marine ecosystems due to human activities, entering principally through wastewater. Nevertheless, there are major knowledge gaps about their environmental impact, behaviour, effects, and their potential to bioaccumulate and biomagnify through the different trophic levels (Álvarez-Muñoz *et al.*, 2016) (Figure 11.2).

The presence of these substances in marine environments can be studied from different perspectives. Environmental science disciplines, such as ecotoxicology, investigate the effects of contaminant exposure on all organisms in the biosphere, including humans. However, it is difficult to determine how the impact of a chemical substance can interact at a molecular level and propagate at population and community levels. Furthermore, there is a lack of understanding concerning the complexity of contaminant mixtures or "chemical cocktails" and their interaction with climate change stressors such as temperature, salinity and acidification (Stauber, Chariton and Apte, 2016; Kaushal *et al.*, 2018). In social science and humanities, there is an increasing interest in analysing the relationship between chemical substances and socio-natures, referred as the "chemical turn". From a chemical perspective, questions have arisen about the ontology and metabolism of these substances between the living and more-than-living (Romero *et al.*, 2017; Balayannis and Garnett, 2020). Chemical geographies gather different

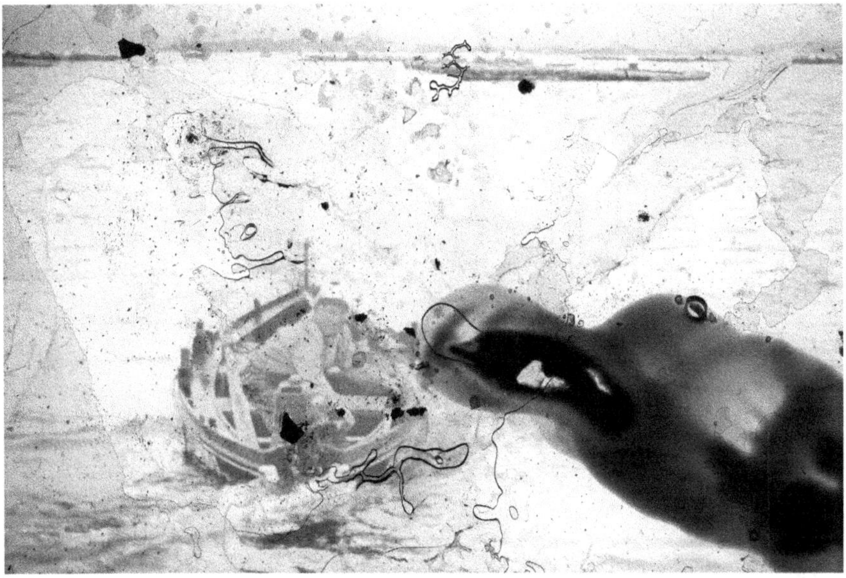

Figure 11.3 Altered photograph from Gonzalo Iglesias's personal photo archive, 2018.

disciplines that share the common interest of thinking through synthetic chemicals: how these substances are embedded in the contemporary world (Romero *et al.*, 2017) (Figure 11.3).

In an attempt to imagine the chemical geographies of the Anthropocene, we use the seascape to approximate evocations that the sea continuously awakens. The notion of seascape is broad and ranges from concrete conceptualizations in fields such as geographic sciences (e.g. spatial planning) to other more abstract elucidations, which understand seascape as a representation of the ideas we have constructed about the sea (Pittman, Kneib and Simenstad, 2011; Ingersoll, 2016). Our approach to the seascape concept brings into dialogue two contrasting stances. On one side, an anthropocentric perception sustained by human experiences and its subsequently image production, in contraposition with the less perceptible relationships between chemicals and marine ecologies: the subterfuge of what happens on a non-perceptible molecular level. We draw near the debates about the Anthropocene that moves beyond its original conceptualization from physical sciences, towards notions such as Capitalocene or Chthulucene that incorporate a myriad of capitalism in the web of life and the relevance of tentacular thinking in times of living and dying together (Haraway, 2016; Moore, 2017) (Figure 11.4).

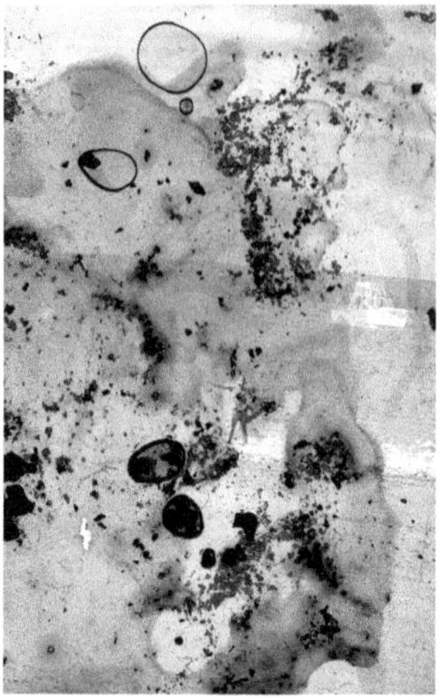

Figure 11.4 Altered photograph from Oscar Jara's personal photo archive, 2018.

A wide variety of blasted landscapes are spreading across the globe, affecting relations of life in diverse ways (Tsing, 2014). Degraded landscapes are unevenly distributed, as a consequence of the externalization of costs to sustain neoliberal lifestyles (Murphy, 2012). Pollution is one face of the geometries of blasted landscapes. Synthetic chemicals have penetrated everything. The contamination that confronts us is disruptive. On the other hand, contamination binds us, it opens pathways to sustain other possible encounters and belongings (Tsing, 2015). While it is crucial to defy and resist pollution, it is also meaningful to overcome troubled notions of purity. As Shotwell indicates, "purity practices – in ideology, in theory, in practice – work to delineate an inside and an outside; they are practices of defining a 'we'" (Shotwell, 2016, 13). Purity practices produce an idea of normativity, delimiting what is good and desired and what is not, which reinforce separation and unevenness (Shotwell, 2016). Hence, it is not about accepting contamination and its consequences, it is recognizing that we live in a "permanently polluted world", with the contradictions and challenges it embraces (Liboiron, Tironi and Calvillo, 2018: 343) (Figure 11.5).

Engaging with the exercise of exploring how chemicals are embodied in seascapes we inquire the following questions: what do chemicals represent beyond their physical-chemical expression? In which way does the phantasmagoria enable us to understand the current chemical age and to re-imagine the Anthropocene? This project responds to a collaboration between arts, political ecology and critical geographies, pursuing multiple stories and relationships between chemicals and waterscapes in the production of

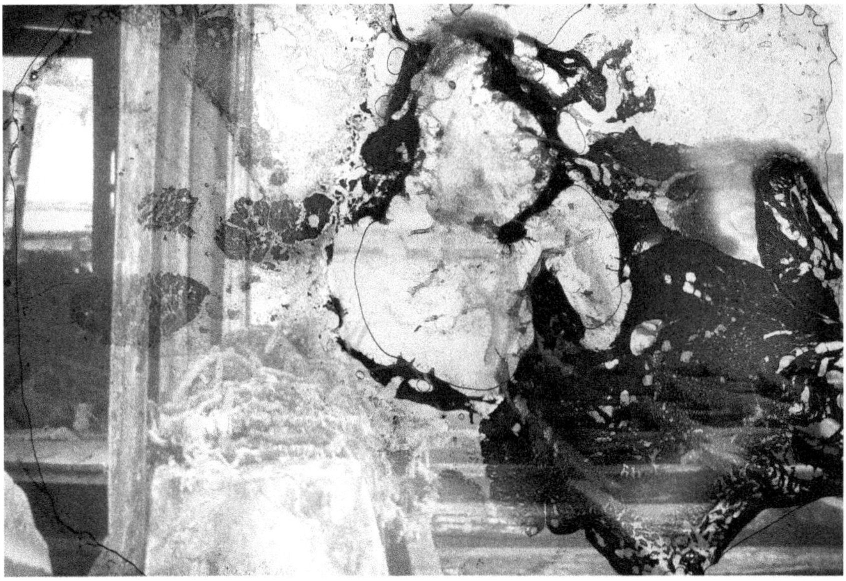

Figure 11.5 Altered photograph from Gonzalo Iglesias's personal photo archive, 2018.

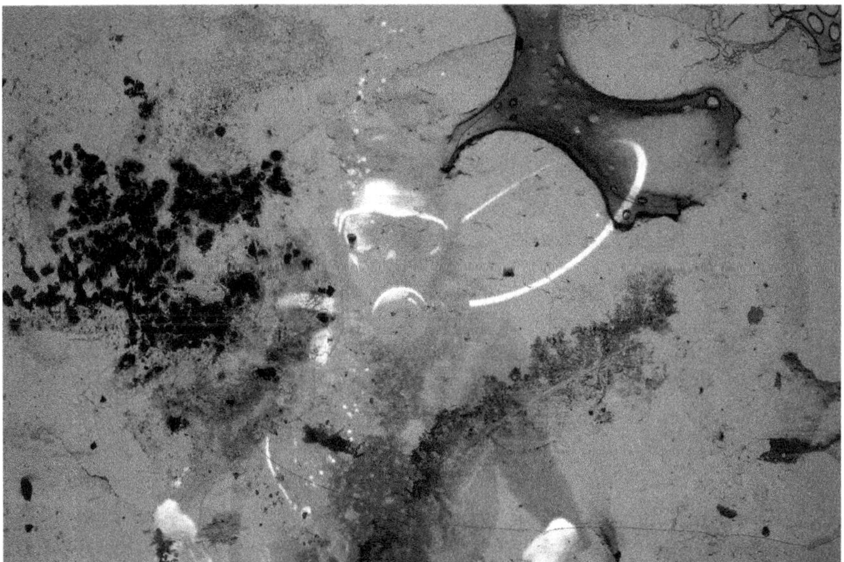

Figure 11.6 Altered photograph from Soledad Castro's personal photo archive, 2018.

Figure 11.7 Altered photograph from Laia d'Armengol's personal photo archive, 2018.

socio-natures. In the active interplay contaminants have with sentient and non-sentient parts of marine ecosystems, sea stories remain mysterious and unanswered. CECs create dystopias not imagined before, of a seascape haunted by our own ghosts (Figures 11.6 and 11.7).

About spectres and phantasmagorias

A central part of our project deals with fear. Fear and discovery of something that haunts us, that approaches and pushes us towards death. It is the recognition that something is not right. We cannot grasp it because it is elusive, intermittent and indeterminate. However, we know of its presence; in its furtive nature it leaves traces and vestiges of its appearance.

What are we afraid of? Death, the end of the world, the apocalypse. Late discovery of occasioned destruction. We remember the image of Walter Benjamin's angel, with constant destruction appearing before his feet, in the unstoppable flight driven by the wind of progress (Benjamin, 2002). On his way to the future, that we imagine with terror, this angel has realized that there is no further path, he has destroyed everything. Everything we produce, consume and use reappears. We sweep the dirt under the carpet and it appears in our throat (Figure 11.8).

In this gloomy scenario, we must converse with the spectres that arise. The spectres that appear from this destruction can teach us to live better.

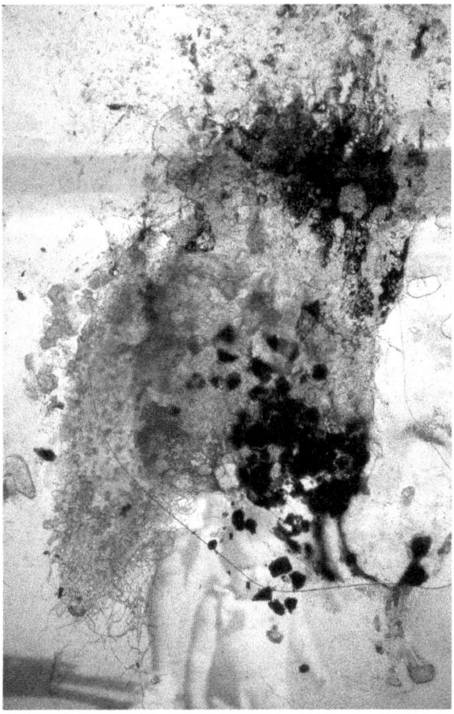

Figure 11.8 Altered photograph from Daniela Mora's personal photo archive, 2018.

"If it—learning to live—remains to be done, it can happen only between life and death." (Derrida, 2006: xvii). What happens in this dialogue between both of them, not separately. For this to happen, the intervention of the ghost is necessary. To talk about the ghost and with the ghost, in turbulent times loaded with political and ethical longings, must be driven and based on respect for those who are here, who will arrive and who eventually will die. This learning to live means to live in another way, "not better, more justly" (Derrida, 1995: 12–13). It is then justice that mobilizes us to create an idea of the world that incorporates the dead (future and past). The ghost enables us to think about it. It wanders between the material and the immaterial, between the past and the future, disjointing the relationship between the two. This dislocation between material and immaterial appears as capitalist spectrality, with the commodity fetishism and its abyss between symbolic value and real value (Figure 11.9).

Commodity fetishism, studied by Marx, has been re-appropriated under the figure of the spectre by Derrida and Benjamin. For Benjamin (2002), the phantasmagoria refers to the ideological concealment of capitalist society's commodities. Commons become fetishized when the exchange value is disjointed from its use value, situating the value of the commodity as a

Figure 11.9 Altered photograph from Oscar Jara's personal photo archive, 2018.

spiritualized feature of itself. Therefore, the phantasmagoria shows these supposed fixations between the object and the symbolic meanings. We understand CECs, which are residual elements of consumer goods, as arising traces of the spectre of the commodity. The ghost arises in the midst of this displacement between the material and the symbolic. Its residue is a return to its materiality, it is what remains of the object. This materiality points to the ghost but in its separation, in its carcass. In this way, the ghost is not attached to the object, but corresponds to it. Just as when we observe a corpse, we cannot think about it without its previous animation, we cannot see the contaminants without its phantasmagoria. This material carcass of the spectre in turn functions as a remainder upon death (Figure 11.10).

Apparitions from the Anthropocene: artistic procedure

The artistic procedure consists of a photographical archive from different people about their experience with the sea. The archive was made in collaboration with persons from distinct latitudes who were invited to respond to the following questions: how does the sea relate to your life? How could you

Figure 11.10 Altered photograph from Soledad Castro's personal photo archive, 2018.

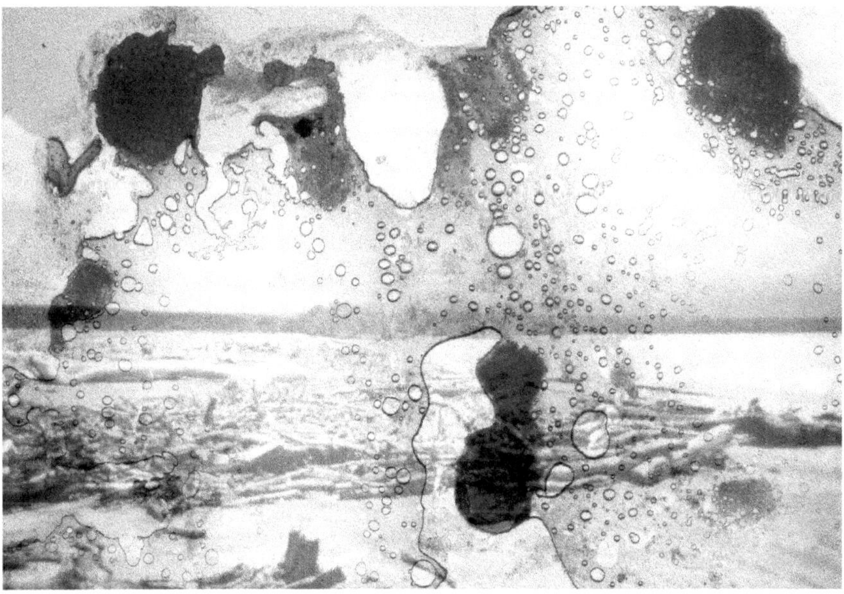

Figure 11.11 Altered photograph from Sara Granata's personal photo archive, 2018.

synthesize in an image your relationship with the sea? The answer to these questions resulted in a series of digital photos, in some cases with little personal stories attached to them. It became crucial to convert the archive into physical or material images. Hence, we transformed the virtual photographs onto slide-film, which is a positive image in a transparent base. Afterwards, we altered the materiality of the film by using various sources of CECs commonly employed in homes, such as coffee, sunscreens, pharmaceuticals and fragrances, among others. In this procedure, we display the materiality of the pictures with the traces of CECs as a way to interweave contaminants with our symbolic construction upon the sea. The result was presented as a lecture performance, where the films appeared and disappeared through time by using a slide projector as a phantasmagoria apparatus[1]. The final knot consisted of returning the altered virtual images to the involved participants, to bind the discourse between CECs, personal memories and blasted seascapes (Figure 11.11).

Conclusions

The phantasmagoria describes the apparent fixation of cultural products with their symbolic meanings (Benjamin, 2002). The contemporary chemical merchandise functions as ways of solving health problems or enhancing productivity. These solutions become the spiritual value behind the consumer products, which, although functional, also reveal a character of "semantic surplus value". They offer a chemically improved lifestyle in a

Figure 11.12 Altered photograph from Sara Granata's personal archive, 2018.

society whose end is emptied in its accelerated rhythm of production and consumption. For Benjamin, phantasmagoria apparatuses project images over the real world, creating an undifferentiated unity. The symbolic surplus is undifferentiated from the materiality of the commodity in CECs, leaving not visible but existing polluted traces behind. These contaminants, conceived as necessary in capitalist societies, mark a sign and dictamen of our time, our fetishes and needs. They reveal hidden emotions, desires and ideals – unveiling in their rhythm of production and consumption, the voracity inherent to the Anthropocene era.

The process of destruction and alteration of the photographs seeks not only to visualize the life and dynamism of chemical materials after their use, but also to unfold, through images, the phantasmagoria behind our gaze upon the sea. The appearance of the traces left by chemical spectres mutates and deforms the bucolic anthropocentric images of a pristine sea that has already disappeared, or maybe never existed (Figure 11.12).

Note

1 Phantasmagoria apparatuses were a family of technical objects whose operation consisted in projecting images onto the real world, so that both the real and the projected image were confused, producing horror and astonishment in the eighteenth century.

References

Álvarez-Muñoz, D. *et al.* (2016) 'Chapter 1- Contaminants in the Marine Environment', in Blasco, Julián et al. (eds). Academic Press, pp. 1–34. doi: https://doi.org/10.1016/B978-0-12-803371-5.00001-1.

Arcuri, A. and Hendlin, Y. H. (2019) 'The Chemical Anthropocene: Glyphosate as a Case Study of Pesticide Exposures', *King's Law Journal*, 30(2), pp. 234–253. doi: 10.1080/09615768.2019.1645436.

Balayannis, A. and Garnett, E. (2020) 'Chemical Kinship', *Catalyst: Feminism, Theory, Technoscience*, 6(1), pp. 1–10. doi: 10.28968/cftt.v6i1.33524.

Benjamin, W. (2002) *The Arcades Project*. Cambridge: First Harvard University Press Paperback Edition.

Bernhardt, E. S., Rosi, E. J. and Gessner, M. O. (2017) 'Synthetic Chemicals as Agents of Global Change', *Frontiers in Ecology and the Environment*, 15(2), pp. 84–90. doi: 10.1002/fee.1450.

Derrida, J. (2006) *Specters of Marx The State of the Debt, the Work of Mourning and the New International*. London: Routledge Classics.

Haraway, D. (2016) *Staying with the Trouble*. Durham: Duke University Press. doi: 10.2307/j.ctv11cw25q

Ingersoll, K. A. (2016) *Waves of Knowing*. Durham: Duke University Press. doi: 10.2307/j.ctv125jhnn.

Kaushal, S. S., Gold, A. J., Bernal, S., Johnson, T. A. N., Addy, K., Burgin, A.,... & Belt, K. T. (2018). 'Watershed 'Chemical Cocktails': Forming Novel Elemental Combinations in Anthropocene Fresh Waters', *Biogeochemistry*, 141(3), pp. 281–305.

Liboiron, M., Tironi, M. and Calvillo, N. (2018) 'Toxic Politics: Acting in a Permanently Polluted World', *Social Studies of Science*, 48(3), pp. 331–349. doi: 10.1177/0306312718783087.

Moore, J. W. (2017) 'The Capitalocene, Part I: On the Nature and Origins of Our Ecological Crisis', *Journal of Peasant Studies*, 44(3), pp. 594–630. doi: 10.1080/03066150.2016.1235036.

Murphy, M. (2012) 'Chemical Regimes of Living', *Environmental History*, 13(4), pp. 695–703.

Naidu, R., Jit, J., Kennedy, B. and Arias, V. (2016) 'Emerging Contaminant Uncertainties and Policy: The Chicken or the Egg Conundrum', *Chemosphere*, 154, pp. 385–390. doi: 10.1016/j.chemosphere.2016.03.110.

Pittman, S. J., Kneib, R. T. and Simenstad, C. A. (2011) 'Practicing Coastal Seascape Ecology', *Marine Ecology Progress Series*, 427, pp. 187–190. Available at: www.jstor.org/stable/24874687.

Romero, A. M., Guthman, J., Galt, R. E., Huber, M., Mansfield, B. and Sawyer, S. (2017). 'Chemical geographies', *GeoHumanities*, 3(1), pp. 158–177.

Shotwell, A. (2016) *Against Purity: Living Ethically in Compromised Times*. Minneapolis: University of Minnesota Press.

Stauber, J. L., Chariton, A. and Apte, S. (2016) *Chapter 10- Global Change, Marine Ecotoxicology*. Cambridge, MA: Elsevier Inc. doi: 10.1016/B978-0-12-803371-5.00010-2.

Tsing, A. L. (2014) Blasted Landscapes (and the Gentle Arts of Mushroom Picking), in *The Multispecies Salon*, pp. 87–109. doi: 10.1215/9780822376989-005.

Tsing, A. L. (2015) *The Mushroom at the End of the World*. Princeton: Princeton University Press. doi: 10.2307/j.ctvc77bcc.

12 Study for "Memories of the apocalypse"

Carl Christian Olsson

Study for "Memories of the apocalypse"

We have never stopped repeating that becoming is a dangerous affair, although there is much more to say about the risks and failures of plan(e) making than we have so far. One must be ever careful lest the weight of the strata will come crashing down or otherwise slip away like dust and silt. Sometimes one wakes up, passes through a door and is swept along on an unexpected journey. Perhaps one takes up driving after a long time. And sometimes there is a risk of being led toward disaster, or else disaster seems unavoidable as the plane of consistency is drawn too fast and revealed for what it can become. A car crash is a haecceity – a fleeting moment between two questions: "What happened?" "What will happen?"

Since we insist on the reality of becomings, we must insist on the reality of their attendant dangers. That the perils were always there is often apparent only after a disaster transpires or else they are perceived but deemed to be inconsequential. And yet Ahab does not chase the white whale alone. Although he may convince his crew to follow him through the white wall their families and friends are left behind on dry land, wondering and grieving – left behind knowing nothing of that mad becoming-whale.

Revelations occur when secrets seep into the blinding light of day or are covered by the folds of night. Two moments at which one can say: "No more subterfuge. No more shadow". Midday, and Zarathustra's laughter is resounding throughout the diatremes of Earth. Something has happened, and at the same time the fact of this something is revealed for what it is. A revelation is the ruination of the strata – what remains of them once their most intimate secrets have been uncovered. To the full body without organs we have already added its empty and cancerous counterparts. Now we can say that there are at least two dangers facing becomings-animal, becomings-woman and becomings-imperceptible. Either the fugue lines become lethal and turn to self-abolition and utter ruination or else they turn on themselves, bounce back into black holes with very different consequences compared to the former case. All of this is a vast oversimplification. A more accurate description must consider the concrete effects of the abstract machines in question.

DOI: 10.4324/9781003128854-13

The question is not if the machines are abstract, since they clearly are. The question is if they can be disentangled from the concrete assemblages they effect and the planes they draw.

Diseases, virulent infections and plagues can be considered through their affects and means of propagation; what they can do and the speed at which they travel. Spread in secret, an infectious disease remains hidden in a population until it suddenly strikes with incapacitating consequences. The first victim of an illness is normally ignored, even after his or her symptoms are obvious for all. This victim, or stranger, is a dark harbinger of what is to come and is thus anteceding the arrival of the plague itself by probing new lines of transmission.[1] After the initial odd cases of now-sick strangers arriving healthy at the city gates, or of pale figures stalking the hinterlands, events accelerate. The coming of the plague is an event. It is nevertheless possible to provisionally enumerate at least three principles belonging to the plague-assemblage as long as we understand that such a list is not to be considered exhaustive.

Double reciprocity

A stratoanalytic approach to the bubonic plague should begin with a symptomatological inventory of its effects on the body. The distinctive pustules associated with the disease are zones of combat between the affected and unaffected parts of the body, but this does not entail a concentric propagation from one bubo to the next. In this situation the arborescent lymphatic and cardiovascular systems become the staging grounds for the appearance of new pustules in quite different locations from one another. Moreover, the most common localization of the outbreaks to the armpits and inguen is indicative of the vampiric tendency of bacteria like *Yersinia pestis* upon the lymphatic system without which the disease could not progress. The corporeal combat is correlated with a similarly distributed tensile metamorphosis of the mind.[2] Sometimes, the noetic content of the plague is expressed in the form of a feverish delirium, bordering hysteria or a disjointed dream. Sometimes, it takes the shape of apathetic resignment before the inevitability of death. Between such extreme attitudes, the veritable battle that is the plague is enacted. But despite the specificity of these phenomena, it is evident that plague is not limited to effecting passions of the body and mind at the individual level, but constitutes a molecular or particulate liberation pertaining to haecceities. The plague enters into an assemblage with the body and despite modifying the organism more or less predictably does not lead it in any determinate direction, and rather carries it every which way in an aberrant and unpredictable manner. This can be transposed to the social stratum where plague-stricken regions are often home to mad festivities and spontaneous carnivals testifying to the transmogrification whereby collective behavior is put at the service of the plague as if to celebrate and welcome its ordeals.

Cartography

It used to be held that no epidemic ever struck by chance. While some diseases are indiscriminate in scope and make everyone their potential victims, others remain highly selective and more ruthless in certain geographical areas. The basis according to which a selection is made is itself highly variable, although there are those writers who have made guesses about its operational basis. Take the case of Procopius who provides a first-hand account of the Plague of Justinian, an affliction that "seemed to move by fixed arrangement, and to tarry for a specified time in each country, casting its blight slightingly upon none, but spreading in either direction right out to the ends of the world, as if fearing lest some corner of the earth might escape it".[3] It is crucial to recall that prior to Pasteur's microbial revolution, epidemics were entirely religious or mystical events. The dynamisms of disease thus conceived are, however, poorly understood if they are interpreted in the language of divine retribution. As Procopius attested, it seemed as if only the most base and reprehensible remained unaffected precisely because they were forced to don masks of false benevolence. In other words, Procopius' account of the plague is characterized by a certain suspension of bad faith, but this is itself a misguided notion. Although all would have their due share of pestilence, this share can only be articulated with reference to an aleatory principle not unlike Lucretius' swerving atom. The plague is an altogether foreign force that makes a mockery of the very terms of moral justice to the benefit of a very different selection that can only be understood from the perspective of a cartography capable of accounting for both the extension and intension of the disease.

Alliance and Segmentation

Plagues are accompanied by conspiracies that multiply without reason, apparent revelations that pop up here and there and rumors about coming death. And when the plague finally arrives, its secrets are no longer a private affair as the entire world is delivered to delirium. That disasters are capable of transforming societies is well known and it is not surprising that migration and nomadicism have been associated with the spread of disease as a cutting edge of deterritorialization. José María Arguedas describes an epidemic that arrives from afar, striking terror in all peoples.[4]Before the approaching plague, Indians and Spanish; colonized and colonizer are alike. Plague stricken becomings. In all this, the first principle that we enumerated must not be forgotten: a becoming-intense proper to the plague's unique capacity for reciprocity. A pure intensity of disease that is no longer reducible to explanation on microbial grounds. Instead the plague implies relations of movement and rest, relative speeds and slownesses alongside and against more or less developed strata. Entire societies have collapsed when their components have started scuttling and dancing across an eroding ground. It is an unfortunate fact that there are those paranoiacs who seek to mete out responsibility, always asking about who is to blame – "who sent these agues and buboes to us, why now and here?" That this way of thinking is not limited to the language of divine providence is shown by Foucault's analysis of the segmentation of the plague-stricken city. Although it is of interest in its own right, the same paranoiac model that characterized the city's response to the plague can be generalized across contexts and it is even suggested that the prison and delinquency (the form of content and form of expression) proper to disciplinary power developed from an increase of society's degree of organization in response to the plague.[5] It would be a serious mistake to conclude from this that nomads and migrants can be construed as allies of the plague and, as Foucault shows, there exists an entire history of how the state has taken advantage of the spread of disease. There is no deterritorialization without the possibility of reterritorialization and a simultaneous development of more rigid forms of segmentarity and disciplinary techniques. That rumors, conspiratorial ideation and xenophobia are common responses to the plague further belies the moral ambivalence of contagion and its simultaneous uptake in systems of cruelty and doctrines of judgment.

Judging by this case we were perhaps wrong or vastly oversimplifying things when we said that rhizomes can be associated with cartography and with maps. (This is not to say that revelations brought about by abstract machines are limited to contagions, even if the latter may be a propaedeutic case.) What we must say in addition is that even cartographers can fall back onto the strata – are just as complicit in drawing rigid lines, are just as prone to fail at constructing the plane as anyone else and are just as likely to climb trees and survey their territories from above. To the cross, the gavel and the couch we must add the sextant. Cartographers, no less than other disciples of God, can speak according to the terms of a transcendent judge. In order to understand this, it is imperative to reconsider the misfortunes that can occur in the passages from one plane to the other and return to the stakes of plan(e)making.

It is clear that there were always risks tied to becomings and it can even be said that becomings were only ever exercises in managing incalculable and necessary risk, not to just oneself, but to the assemblages one enters into. Perhaps risk constitutes the relationship between becomings and an apocalyptic sensibility that sometimes connects to images of deserted landscapes and ruins and sometimes leaps into a musical register. Did Messiaen's Quartet make it possible to reappropriate the Nazi war machine if only for a fleeting moment? Sometimes there are less morbid cases, such as in Woolf's novel where a man travels to Constantinople, returns as a woman and leaves the great city on its path to decline. Talk of ruins and apocalypse has always been a question of the extreme points of becoming, but the association of revelation with eschatology in the New Testament precisely reverses this relationship and subordinates revelation to an expectation of destruction as a matter of interiority or analytic necessity.

The Gnostics believed in the existence of a divine spark that had been imprisoned in their bodies. In order to free their souls from this corporeal imprisonment, they required a higher knowledge that could be attained through certain spells and incantations. We are not to be mistaken for affirming either the eschatological or the Gnostic view according to which the strata must be purged, but still, can one sorcerer not borrow a little magic from another? Is there not a certain superiority of the plane of consistency even if it is never sufficient for salvation? Perhaps as long as it is merely a matter of freeing oneself from the strata that are always purporting to impose their transcendence. The whole point of what we have called pragmatics is to draw another plane for that which is no longer oneself, but this is never the same as to have done with this world. The point is to have taken a chance and to have risked shattering the world into a thousand inseparable pieces, each of which can be the key to returning to the forms and subjects on the plane of organization. Moreover, the two planes are always drawn together, perpetually birthing one another and never without repercussions. And between them other planes are always made and unmade. Whenever we have warned of the dangers facing the plan(e)maker, we have been careful not to say that disaster is the inevitable result of becomings-animal, becomings-molecular and becomings-imperceptible, that every becoming ends in a violent explosion or catatonic collapse, either on an empty or cancerous body without organs; but it is time we entertained this possibility. To dissolve the strata or wrest free of their grasp is a dangerous business. Is it nearly enough to say that the risks are ever-present and cannot under any circumstances be eliminated? For is there ever a chance of something like lasting success beyond the threshold? Is ever the plane of consistency grasped as more than a fleeting moment between its fates in self-abolition and catastrophe of return? Would this mean that becomings-animal, woman, molecular or imperceptible are never to be attempted or only that one must be ever more vigilant? The last question itself suggests that there is such a choice to be made.

In either case we cannot let the well-worn names of Artaud, Nietzsche, Lawrence and Fitzgerald fool us into believing that we will be saved, that toying with so many Rhinoceroses' horns will be enough, that multiplication will leave us a little better off than before, that a sip of water will redeem us, make us free from judgment and in possession of our power (*pouvoir*). We are saying that much more is at stake, that joy is always permeated by a terrible cruelty, soaked in it through and through. Beyond Professor Challenger and Randolph Carter and beyond even the accursed Ahab there have existed yet more unfortunate souls. An old violist is swept away like a bird carried by a foul wind.

It was more horrible than anything I had ever overheard, because I could now see the expression of his face, and could realise that this time the motive was stark fear. He was trying to make a noise; to ward something off or drown something out – what, I could not imagine, awesome though I felt it must be. The playing grew fantastic, delirious, and hysterical, yet kept to the last the qualities of supreme genius which I knew this strange old man possessed.[6]

A window that is also a threshold opens and the old man is no more. In his place are only speeds and slownesses, movement and rest – an "unimagined space alive with motion and music" without "semblance to anything on earth". All around "him", now caught up in a becoming-bird, lie sheets of music or upturned leaves. It is no coincidence that this is a dream of the schizoid Lovecraft, "the *least* sensuous of all living beings… tending to view and enjoy all things as a passive, detached, and sometimes remote spectator".[7] Behind every schizoid disposition of this kind there is pent-up schizophrenia waiting to be unleashed. And when the moment comes, everything is staked.

Every deliriumis oriented toward the world and runs off in all directions – Schreber and Rimbaud, Lovecraft and Lawrence. Each delirium takes into its grasp races, peoples and cultures – their becomings are indeed constituted by so many delirious racial machines, peopling machines and cultural machines; machines that deride and chastise as often as they laugh and cheer. It is clear that this has nothing to do with Kraepelin's and Bleuler's conception of schizophrenia and its supposed association with autism, which is only a consequence of the disruption in the clinical setting of a journey or machinic process on another plane.

It should be noted that the hierarchical picture that this talk of planes gives rise to is nothing other than a misconception and can only be a figure of speech that takes us so far. As we have said, committing to such an account would be to misunderstand the plane of consistency and the abstract machines, neither of which contain references to additional dimensions, and pertain to no forms or accounts of formation. At its limit there is not even the constitution and reconstitution of this or that assemblage; only longitudes and latitudes; blockages and flows. Degrees of delirium at noon.

None of this is to stop us from understanding the profound ambiguity of believing oneself to have crossed the threshold or have done with the judgment God. On the one hand, there is the initial struggle against the great strata, the lines of flight and the unnatural participations, the alliances that have nothing to do with the admissible lines of descent. But on the other hand, there is the reversal after which the struggle against the strata may become a struggle on behalf of a new center that coalesces around the subject that replaces the eschewed God by exhausting its own relations, occupying every identity at once. "I, Antonin Artaud, ammy son, my father, my mother, and myself".[8] A new God that will not tolerate your judgment any more than you tolerated His. It is not man that threatens to replace God. It is above all the tyranny of the Self (*Moi*) which risks taking His place even if this Self emerged as the sole repose following a painstaking struggle against the pretensions of priests and psychiatrists. It is because of this horizon that construing plane of consistency as a plane of formation in contrast to a plane of forms would amount to a severe misunderstanding. For the return to the subjective stratum via a solipsistic right would only serve to ossify the content of the racial, sexual and cultural deliria that first dissolved it. Is this perhaps the entire truth of becoming? A perverse revelation after which one returns to a new fixation instead of remaining perpetually imperceptible, carried like another leaf by the wind...

It is true that sometimes a becoming is too violent, an unnatural participation much more than a body can take. That which forces thought to think also breaks the spine as one is taken up by forces that cannot be controlled, under whose influence one is inadmissibly folded, unfolded, refolded. The plague is not the only case even if it was informative. When Joan of Arc was burned at the stake at her tribunal's orders, it was really a second or corollary (not implying the presence of a sequence or succession) immolation since she had already left the plane of her organism long ago. If she felt content before her fate it was only because she had already been taken up in all manners of strange becomings and because the body that was thrice burned was already a mere husk. Rarely has there been a misunderstanding of such grave proportions. For "you can and must conclude that these revelations and visions, if Joan ever had them, proceed rather from evil and lying spirits than from good".[9] The tribunal could only see and speak in order to distinguish between the good or bad natures of revelation and never did they comprehend that it was only ever a question of becoming and a more profound cruelty. *But even if they had understood this they would have been faced with difficulties. For becomings would pose no less of a threat had these not so often been misconstrued in the moral language of priests, psychoanalysts and professional cartographers.* Behind the veil of morality and facetious framing of the problem there remains a secret jurisprudential matter of real gravity. Beset by her revelations, Joan had traced a trail of blood and fire from Orléans to Reims even if her nascent war machine was appropriated by the strategic interests of a monarch in order to reclaim his realm. It is not nearly enough to recall the personal horrors that those taken up in becomings sometimes happen upon. The stakes are much higher and concern entire towns and cities, peoples and communities obedient to the law of the land, but of apparent innocence.

Becoming is ignorant of laws and customs. It follows a combinatorial game of chance to test the limits of bodies, to discover under what circumstances they can endure what affects. We saw that priests have reversed the meaning of revelation, which is to say they have never grasped becomings. For the English clergy, Joan was precisely in communion with the devil, the powers of whom they could only understand in opposition to the laws of God. What was Cauchon to do when he was presented with an enemy of the law, all the more insidious because of her heretical claims to have acted out of divine inspiration? In fact, Joan's revelations were demonic in a quite different sense. What the tribunal could not understand was that the question of good and evil had ceased mattering long ago, or maybe it never did in the first place; that the source and moral standing of Joan's revelations were strictly speaking secondary to their intensity of her singularities and of the power of the assemblage which she had been taken up in by being carried along. On the plane she had drawn in her encounter, laws, customs and even life and death no longer seemed to matter – there were only movements and speeds; another kind of multiplicities. When Geoffroy said that combustion is a process that releases organisms from their organization, it was no doubt to make a general point about destratification. While combustion is indeed characterized by a release of smaller particles, it is also a becoming-molecular of the molar aggregate so that it is never just the body on the organic stratum or the Self (*Moi*) on the subjective stratum that combusts or passes away in combustion. We can now see that the way in which destruction can be an inevitable outcome of becoming is only intelligible in relation to its concrete instantiation in a way that forecloses any universal answers.

This is why psychoanalysts, like priests, have never understood the nature of revelation. The death instinct is the exemplary case. Freud would have us believe that the organism is subjected to a concrete, though unknowable drive toward death, which he went on to extend through the social machine, thereby effecting the demise of desire, and consequently of revelation. The inversion of revelation in the anticipation of an "apocalypse" that we already saw in the case of Christianity is only repeated by psychoanalysis. This is because the death instinct inextricably links incalculable risk with a drive toward the inevitability of death and destruction. The problem is precisely the reification of Thanatos as a separate instinct, somehow distinct from Eros. The problem is not death per se. The problem is how death is inscribed within primary process itself. Sabina Spielrein did not know how to have done with this theatrical organon that shackled the unconscious and its abstract machines and it was precisely this misstep that enabled Freud to draw his ridiculous conclusions.[10] Psychoanalysts have never understood that death and disaster are both to be conceived in relation to a wild dance at noon, a veritable festival of cruelty during which disruption is only one of many concrete possibilities. Neither have they appreciated that one is always already upon this stage, that one occupies its edges in a continuous sliding off of it to the hum of one or a thousand tiny machines. Instead the creative and destructive tendencies were amalgamated and opposed; one always following from the other when in fact it is never a problem of sequence and implication, but of an open series in which creation and destruction only mean continuation and disruption. Contrary to the chronic fixation on the mutual implication of life with death in psychoanalysis, there is the Proustian schema with its fleeting time, the consequences of which are by no means less troubling than what was implied in the psychoanalytic circumscription of the Libido. The name for this death is indeed catastrophe or apocalypse, but it is always an accident that can occur on the far side of a white wall or threshold even in those cases where the accident is a matter of concrete inevitability on a line of abolition. For death is only ever one of many extrinsic outcomes and can never be reduced or assimilated to an instinctual drive.

In one of J. G. Ballard's works, a character goes on a journey, a becoming-imperceptible, becomes a fleeting series of images that is itself a sort of perpetual dying. From the viewpoint of the strata, something escapes through an epistratum into a newly associated milieu which is folded back onto the strata leading to the demise of both, in the middle of which it may seem as if "the planes of his face failed to intersect, as if their true resolution took place in some as yet invisible dimension, or required elements other than those provided by his own character and musculature".[11] The build-up to the disaster was also described by Van Gogh as an iron wall that separated him from what he could do: a wall that must be undermined at all costs. His realization that this could only be done with the utmost care and effort testifies to his practical ingenuity. The artist's trajectory demonstrates how difficult this can be, that any moment can be a failure of the requisite patience; from the drab palette of the early works to the sudden outburst of color for which everything would be chanced in an unstoppable march toward a final moment. Was it worth it? Perhaps if one day a painter comes along who is capable of forming an assemblage to the effect of rendering again the lines and commas of Van Gogh's face, for whom the world will appear as a delirium in which so many features of the landscape appear folded onto so many faces. At work in these paintings will be all kinds of abstract machines that deterritorialize the visage only to discover new milieus for its traits to reterritorialize upon, such as sea urchins and pies. Meanwhile, the no longer human figure will enter a becoming-imperceptible and meld with the ocean waves, with interiors and indistinct features crossed by abstract lines that seem to inevitably end in nuclear blasts. There would no longer be faces or landscapes, nor even a close-up of faces that are also landscapes nor of landscapes that are also faces – only molecular facility traits and landscape traits resembling nothing, perpetually drifting across the surface of the Earth.

From Diogenes to Artaud, the task of every dog has only ever been to subvert the image of those responsibilities that form the face into a system of black holes and white walls and substitute for them an empty ground beyond ground as the source of all abilities that are attainable by means of an ostensibly irresponsible behavior. But even the great Cynic sage was a little puppy, as domesticated in fact as he was wild in appearance; ever the faithful hound of Antisthenes and the licensed fool of Athens. We contrast him with the life and reign of Heliogabalus, the accursed emperor. Crowned at the age of 14, the incompetence of this child in the eyes of the empire, his incessant debauchery and subsequent assassination by the Praetorian guard. Heliogabalus was all the paradoxes of an anarchic spirit enthroned, seeking and failing to subvert all that was Roman from within, a real contradiction in terms. To crown and raise to eminence one without rule will end in disaster; a double suicide by virtue of unstoppable force: the crisis of the empire and the violent murder of the boy emperor by his own subjects. Crowned anarchy or a rule with only criteria makes a mockery of the law. Such a situation is one where a minoritation element is posited as (1) an object of supreme importance erected within an order as its grounding principle or keystone and (2) itself an agent of "disorder" at the heart this order. This is not to say that the crowned anarchist avoided the subjectivism that plagues representations of a ground, but neither is it a question of dialectic development. On the contrary, crowned anarchy does nothing but amplify subjectivism until it becomes solipsism; until the stratum becomes so rigid that it collapses under its own weight. Through his excesses the opulent emperor would stage the collapse of himself and the state apparatus with it. He brought all of Rome into an apocalyptic destratification that his subjects were not prepared to depart upon. Heliogabalus was not content to destroy himself, he insisted on following a line of abolition so thorough that it would make the empire's eventual disaster inevitable and for that he had to die, cut down in the latrines of his own guard.

A revelation is the unstoppable procession of becoming, and more precisely it is the point at which any becoming will proceed whatever its cost may be. Following from this, a certain practical principle of becoming can be formulated such that if there is no desert in sight, then one is created by salting the earth; if a body does not accord with the affects you want, then it is beaten into form. What secrets will Ahab discover at the bottom of the ocean once he has broken through the white wall of blubber, flesh and brine? One thing faces us with the perfect transparency of a much calmer sea than Ahab sailed, that Moby-Dick never had much of a choice but to offer up his traits – but did the captain choose? Perhaps not. If it was not he who flew, but a different multiplicity, a molecule, a whale-whaler-ocean assemblage which drifted toward its fate, was it not Ahab himself who first said "Yes" to chance?

Artaud, more so than anyone else, would have us believe that there is a rightful way to struggle against the judgments of God, to refuse them and to enact a transfer of a fascinated Self (*Moi*) onto the deserted Earth inscribed by who knows what signs. That is to say that there is hope, if only because even black cows stir in the night. Artaud never stopped repeating that cruelty is not unbridled anarchy, that there is a method to cruelty and that cruelty is in fact only implacable rigor in disguise. Mere intellectual intuition may no longer be an option to reconcile the organized image of the Earth above – drawn and quartered by sextant and gavel at the hands of cartographers and priests – with the plane beneath. Despite the risks Artaud never stopped calling for an alchemical sensibility, like Nietzsche never stopped decrying the charlatans and peddlers of base values, like Lawrence never stopped staggering through his desert on the other side of the Earth. What does it take to learn what a body can do? For the champions of cruelty secrecy and witchcraft certainly exist as means of flight.

But there are many kinds of secrets abroad in the world, and there are cases where one's route through sorcery is barred. Frantz Fanon spoke of secrets, but ones that would be stolen from him unjustly. Like a magician, I robbed the white man of 'a certain world', forever after lost to him and his. When that happened, the white man must have been rocked backward by a force that he could not identify, so little used as he is to such reactions. Somewhere beyond the objective world of farms and banana trees and rubber trees, I had subtly brought the real world into being... The white man had the anguished feeling that I was escaping from him and that I was taking something with me... Everywhere he found only the obvious. So it was obvious that I had a secret.[12] Even secrets can be stolen, revealed or diminished by those sad men who do not understand singularities, nor the intensity of an hour in the sun and who will inevitably accuse others of having robbed them of their worlds; and yet, for a brief moment one can cross a threshold on a line of flight away from such bad faith toward ever more uncertain vistas. In the middle, always in the middle, without an ending in sight.

But if crossing a threshold is such a dangerous affair and death can become inevitable, the question is under what circumstances one can afford to probe one's limits and proceed with the requisite caution. It is by no means given that it is the minorities, those with the "least" to lose who have this chance. This is not about invoking powerlessness in order to sing the lament to the wretched of this earth. Quite to the contrary, what is at stake is a jurisprudential question concerning who can exercise a right to not stake everything in a chance encounter, in other words, to manage their own risk. For there is always the chance that becomings are reduced to a mere game of the bourgeoisie who can dictate their own terms while affording to limit their risk by reallocating it elsewhere: "Michaux and Castaneda; heroes who suffered and experimented". But who, when reading the works of these explorers of the limits of this world, recalls the plight of the Tarahumara?

We, sorcerers and cartographers, are always at peril of forgetting who and what else is taken up in our becomings, who and what we pass by on our flights. We risk forgetting that our travels put entire peoples in danger and should never be taken lightly – and although more careful souls can join our dance, on what terms? We are not presenting any ready-made answers, for becoming is a question of jurisprudence that concerns the creation of novel rights and the invention of new ways of traversing the Earth, but is that in itself not the problem? We are warning that there is, on the one hand, always a possibility that a despot makes a claim to jurisprudence only in order to narrow its scope and bring it to a full stop. And on the other hand that the institution of a new right can have disastrous consequences that are not immediately clear even when the wall is torn down as a final resort. There is an intimate bond between the judge and the cartographer, each of whom can lapse into the other's role at any moment. Even rhizomes can grow into trees that must be pruned unless the entire forest withers; or worse, it may yet become clear that rhizomes were always more insidious than trees. Is this the well-kept secret that is unknown even to those who have found their proper names, those who purport to see the light without noticing the eggshell that lies before their feet? Theirs is the 11th hour *post meridiem*; when noon turns to midnight and Zarathustra cries.

Notes

1 Despite the shortcomings of existentialism, for a fictional arrival of a plague, see Albert Camus, *The Plague* (London: Hamish Hamilton, 1948).

2 Antonin Artaud, "The Theatre and the Plague", in *The Theatre and Its Double*, trans. Mary Caroline Richards (New York: Grove Press, 1958), pp. 15–32. Artaud suggests that not only is the plague a psychic phenomenon, but one that delimits its spread according to an internal logic whose organization cannot be understood with reference to ordinary methods and concepts.

3 Procopius, *History of the Wars: Books I and II*, trans. H. B. Dewing (New York: The MacMillan Co, 1914), Book I, XXII, [7–11]. Procopius also wrote a secret history besides his official accounts of the Persian war.

4 José María Arguedas, *Deep Rivers*, trans. F. H. Barraclough (Long Grove, IL: Waveland Press, 1978).

5 Michel Foucault, *Discipline and Punish*, trans. A. M. Sheridan Smith (New York: Vintage, 1975), pp. 195–198.

6 H. P. Lovecraft, "The Music of Erich Zann", *The National Amateur* vol. 44, no. 4 (March 1922): pp. 38–39, 40.

7 H. P. Lovecraft, "Letter to Virgil Finlay (25 September 1936)", in *Selected Letters V*, 1934–1937, eds. August Derleth and Donald Wandrei (Sauk City, WI: Arkham House, 1976), p. 310.

8 Antonin Artaud, "Here Lies", in *Selected Writings*, ed. Susan Sontag (New York: Farrar, Straus and Giroux, 1976), p. 540.

9 The Trial of Jeanne d'Arc, trans. W. P. Barrett (New York: Gotham House, 1932), p. 183.

10 [TRANS: Sabina Spielrein, "Destruktion als Ursache des Werdens," Jahrbuche für psychoanalytische und psychopathologische Forschung, vol. 1, no. 4 (1912), pp. 465–503. Spielrein served as one of the major sources of inspiration for Freud's attempt to "go beyond" the pleasure principle. Spielrein builds on Nietzsche to discuss the adversarial relationship between creation and destruction within ontogeny and reproduction of the collective.]

11 J. G. Ballard, *The Atrocity Exhibition* (London: Triad/Panther, 1979), p. 8.

12 Frantz Fanon, *Black Skin, White Masks* (London: Pluto Press, 1986), p. 128.

13 Variegated environmental apocalypses

Post-politics, the contestatory, and an eco-precariat manifesto for a radical apocalyptics

Tristan Sturm and Nicholas Ferris Lustig

Perhaps no present has been more future-oriented than our own; and much of this future-oriented thought is characterized by cataclysm. The present age often seems driven by images of its own end. Technological futurists foresee transformative singularities of sudden, stunning evolution to whole new species. Political scientists envision the collapse of states, the descent into lawlessness and violence-plagued anarchy, and islands of order only maintained by vigilant defense and aggressive exclusion. Religious leaders condemn whole cultures and flatter the faithful with prophecies of apocalypse, escape (or Rapture), and vengeance. Heath (social) scientists warn of further viral, bacterial, and fungal pandemics to follow COVID-19, all potentiated by ecological despoliation and extractivism (Lynch, Sturm and Webster 2021; Sturm and Albrecht 2021). Eco-activists illustrate and embellish scientific predictions with warnings of catastrophic climate change flooding major cities, wreaking havoc with intensified storms, shutting down ocean conveyor belts, and driving massive numbers of species into extinction. Radical political theorists mock capitalism for the unsustainable excesses and contradictions driving it toward its downfall. Hollywood movies draw upon each of these and repeatedly rehearse and sell us the different visions of our end. This is an age of variegated and competing apocalypses.

Erik Swyngedouw (2010a, 2013; Wilson and Swyngedouw 2014; Swyngedouw and Ernstson 2018) has put forward the suggestive thesis that such apocalyptic discourses, in particular those that imagine ecological upheaval and climatic catastrophe, have become prime contributors to a post-political condition. This characterization draws upon the work of a series of theorists (e.g. Rancière 2001; Mouffe 2005; Žižek 2009) that describe the present political situation as one in which a managed consensus, populist politics, and technocratic administration have reduced political debates to questions of policy, appropriate technocratic responses, and the negotiation of interests among stakeholders. Real disagreement, dissensus, conflict, and antagonism—the grounds of democracy, egalitarianism, and liberty—have

DOI: 10.4324/9781003128854-14

been increasingly displaced, foreclosed, or eliminated. With the ascendance of neoliberalism, the decline of any alternative competing ideological visions to capitalism, and the lack of an emergence of any revolutionary subject, the current political situation is "structured around the perceived inevitability of capitalism and a market economy as the basic organizational structure of the social and economic order, for which there is no alternative" (Swyngedouw 2010a p. 215). Swyngedouw's original and important contribution to this post-political paradigm is his elaboration of how environmental politics has played a key role in the formation and enforcement of a contemporary post-political consensus. Apocalyptic imagery and rhetoric play a privileged role in this environmental consensus.

In this chapter, we wish to build upon, modify, and, to a lesser extent, critique Swyngedouw's analysis of post-political apocalyptics. Swyngedouw's many writings on post-politics and environmental crisis in this respect act as a foundation to diversify the types of post-political (pre-determined options that disavow alternatives) and depoliticizing (an evacuation of politics) apocalyptic discourses, emphasizing the importance of and differences between what we call contestatory apocalyptics, and advocate for the articulations of seemingly incompatible alliances between progressive contestatory apocalyptics and the varied range of radical apocalyptics that craft fundamental critiques of capitalism and call for emancipatory mobilizations amidst our already unfolding apocalypse. While we agree that efforts to diminish the organization and intensity of democratic and antagonistic forces is a critical tactic of systems of power today, we believe that such efforts should be considered as forms of depoliticization rather than strictly post-political in order to consistently mark its status as an ongoing, partial, incomplete, uneven, and constitutively contested project (Brenner et al. 2010; Darling 2014). Part of this variegated landscape of depoliticization, especially as it relates to apocalyptic discourses, is a variety of depoliticizing gestures other than those accounted for by Swyngedouw. Various religious, scientific, activist, and scholarly discourses mobilize apocalyptic rhetoric to depoliticize environmental issues. We describe a number of different examples and types of these post-political and depoliticizing apocalyptic rhetorics in the second section of this chapter.

We also take issue with how Swyngedouw, consistent with much of the post-political literature, characterizes many contemporary political struggles as not real, authentic, or proper politics but rather merely particularist negotiation of interests that fail to challenge the status quo and help serve to reproduce it. For us, some forms of contemporary contestatory apocalyptic ecological politics serve not only as tools to possibly achieve some immediately progressive gains—a destabilizing of the system, and a cultivation of constituencies for change—but also forming an "impossible political alliance" (Žižek 2009, p. 209). With more radical movements they can serve as preparations and provocations for the emergence of an emancipatory political subject (what we call an "eco-precariat," a term we develop in the conclusion) and more radical (anti-capitalist) transformations of contemporary conditions. The third section of this chapter outlines some of the differences and examples of contemporary contestatory apocalyptics, while the final section sides with Swyngedouw on a call for radical change and takes stock of the apocalyptic discourses that have played and are playing a role in such a pursuit. As such, we conclude with six core principles as a manifesto to thinking about an eco-apocalypse toward building radical politics for the Anthropocene: Earthly, Immanent, Contingent, Conjunctural, Uneven, and Redemptive.

Swyngedouw, critique, and the apocalypse

End-of-the-world scenarios and cataclysmic transformations have caught the attention of a diverse range of contemporary social theory. For us, the two most compelling apocalyptic discourses are those made by radical theorists Erik Swyngedouw (2010a, 2013) and Slavoj Žižek (2009, 2011). Swyngedouw and Žižek are two examples of radical social theorists adopting and adapting apocalyptic imaginations from evangelicals and environmentalists, twisting them into cutting critiques of contemporary capitalist environmentalities and unexpected master signs of a renewed radical insurgence.

Swyngedouw specifies the status-quo maintenance of environmental and apocalyptic issues into a hegemonic "post-political" condition that overlooks the diversity and radical potential within current apocalyptic discourses. According to Swyngedouw (2010a, 2013), a political consensus has taken hold of political imaginations on the question of the magnitude of and proper response to the impending ecological apocalypse. Apocalyptic fears are stoked and then channeled by a manipulative elite politics and a consensual framing that reduces the possibility of a catastrophe to the calming certainty of a manageable crisis; and this crisis response is to be overseen by popularly supported technocratic elites that claim they can solve any problem with solutions generated within the confines of contemporary capitalism. A populist politics that homogenizes humanity in a false universalism makes it appear as if these solutions are being carried out for the sake of everyone and to the benefit of all. No major changes are needed; no major debates need to take place.

James McCarthy, in his article "We Have Never Been 'Post-Political'" (2013), makes several skeptical remarks concerning the validity of Swyngedouw's post-political thesis. McCarthy questions not only the current accuracy of claims depicting present conditions as completely devoid of alternative imaginations and contestations, but also the historical novelty of the post-political. For McCarthy (2013), post-political tactics and tendencies have been an endemic feature of capitalism, and even the use of environmental issues as leverage for depoliticization efforts has been quite common to the history of capitalism. It is important to point out, however, that Wilson and Swyngedouw (2014) in later work do not claim there is anything new about the post-political, but rather concerns around catastrophic climate change have compounded and intensified in the social imagination. McCarthy also questions any characterization of the post-politicization of environmental issues as anything other than a deeply political process by those corporations, scientists, and politicians pursuing such an agenda. While perhaps not the intended message, the naming of the political situation as depleted of political inclusion normalizes depoliticized understandings of contemporary societies. This said, Swyngedouw and Ernstson (2018, p. 20) do not argue that the post-political is a depletion of politics, but rather a subversion of the political in the maintenance of the status quo. McCarthy, as well as Goldstein (2013), points out that Swyngedouw's perpetuation of the post-political also blocks appreciation and even potentially undermines the myriad groups and theorists that have actively struggled to craft systematic critiques of capitalism and to champion visions of attainable radical futures. We feel, however, that this is a mis-reading of Swyngedouw, where on the one hand his reading of the depoliticization is a combined procedure of situating the problem in ecology rather than as political and where on the other hand the apocalypse is deferred to the future rather than recognizing the actually existing, present, and immanent apocalypse for many. This is the geography of it all: the "combined and uneven apocalypse" of the eco-precariat (Williams 2010, p. 11). This twin procedure mobilizes a geopolitics of post-political or techno-managerial solutionism, especially in "the West" in relation to omission of "the Rest," and the pecunious and the precariat.

Swyngedouw (2013), despite the post-political conditions he identifies as rendering alternative imaginations all but impossible or entirely ineffective, articulates a series of arguments meant to position and frame a more radical apocalyptic imaginary. Avoiding what he considers to be the comprised politics of liberal calls for sustainability, and the empty and cynical righteousness of Marxist self-congratulations for predicting capitalism's inability to resolve its contradictions, Swyngedouw (2013, p. 15) claims that a "proper leftist" response should insist on several core principles. First, there must be an insistence that the apocalypse is already unfolding, already impacting the lives of millions in catastrophic, combined, and uneven ways. Second, radical ecological politics should focus on the co-production of humans and nature; the modernist division between them is in part responsible for the callousness of humans toward the environment. A radical and redemptive rethinking of socio-ecologies remains necessary. Third, in contrast to the global humanitarianism of the post-political project, a radical environmentalism must realize that political ecologies are always contested terrains, and even more so in the case of the apocalypse. Adversarial, antagonistic positions must be taken and combative decisions must be made. And lastly, we can only "fully endorse the emancipatory possibilities of apocalyptic life" if we refuse the post-political tactic of marshaling universal means to save the present particular configuration of capital and power. The present particular conjuncture should be used for a radical, universal project of emancipation. We incorporate and expand on these principles toward an eco-precariat manifesto for a radical apocalyptics in the conclusion.

Depoliticization

In this section we seek to both challenge and expand on Swyngedouw's insights into how contemporary environmental apocalyptic discourses are politically disabling. We agree with Swyngedouw that apocalyptic discourses are being deployed to reduce democratic debate and avoid or limit policy alternatives for responses to climate change to the basic ideological coordinates of technocratic administration and neoliberal capitalism. We see such discourses as only one strand of depoliticizing environmental apocalyptic imaginaries (Harper 2019); in other words, the dispositifs of depoliticization—of which the post-political is but one—take a variety of forms and this too is reflected in the variety of apocalyptic discourses: ultra-politics, archae-politics, meta-politics, para-politics (Swyngedouw 2017, p. 57). Besides the scientifically apocalyptic "ecologies of fear" undergirding a spreading managed consensus, a variety of other religious, activist, scientific, and academic discourses are depoliticizing (Diken 2014). Some religious apocalyptics breed passivity by spreading a metaphysical fatalism or embracing an apocalyptic millennialism and some scientists soften their findings in order to dull their critical edge. Here a meta-politics depoliticizes in the archipelago of a "parallel culture amid the interstices of the secular world" (Stephens and Giberson 2011, p. 180).

Swyngedouw writes about apocalyptic discourse but doesn't provide many apocalyptic examples, save for socialism as a redemptive alternative to rapid technological change in the eighteenth century. And while Swyngedouw does contrast religious redemptive apocalypses with ecological ones, he does so largely in passing. He spans historical changes or variances of the post-political which has consequences for how relationships are to be radicalized. For example, Swyngedouw argues that all modern apocalyptic visions lack redemption, realization, or politicization. He writes, "traditional apocalyptic versions still held out the hope for redemption" while new "forms of millennialism around the environmental nexus is of a particular kind that promises neither redemption nor realization" (2010a, pp. 218–219). This schematic division between past and present, religious and secular, is too simplistic. First, "traditional Christian" apocalypses never solely sought redemption. What Swyngedouw does not explore is *how* that redemption came about, whether, for example, through the more meta-political "premillennial" pessimism that posits God's intervention and therefore a heavenly post-political escapism or through the necessarily politicized "postmillennial" optimism that posits human action to bring about the millennium on Earth (Dittmer and Sturm 2010; Skrimshire 2010). Second, the dualistic conflict between corruption of the present and hope for the future is often lost. As the religious studies scholar, Stefan Skrimshire (2013, p. 268), concludes of Swyngedouw, in "ignoring such nuances in apocalyptic narrative, its potential for engaging, provoking, and critiquing ethical and social thought is also being overlooked." Furthermore, the idea that environmental catastrophe is a sign of an imminent apocalypse is not new; rather, such discourses have historically been contemplated within Christianity. Environmental signs of the apocalypse have had a long, perhaps "traditional," history in Christian eschatology from signs produced though the Little Ice Age to famine in the Reformation (Cunningham and Grell 2000). Third, present apocalyptics (as we outline below) have a more varied tapestry of reasonings for depoliticized engagements. And last, the difference Swyngedouw seems to frame in the above quote between religious and secular apocalyptic discourses, while differing as to their conclusions, often find common evidence for catastrophic climate change therefore operating on common registers (Sturm and Albrecht 2021).

Within American Evangelicalism, there are a variety of environmental apocalyptic narratives spanning from more fatalistic American premillennial fundamentalists to leftward Evangelicals of, for example, the US-based but UK-led, Evangelical Environmental Network, which uses their religious voice to promote environmental science and stewardship (McCammack 2008). Many premillennial Evangelicals believe that taking action to reverse this apocalyptic trend of polluting the environment is a challenge to God's nondemocratic and apolitical sovereignty because ecological crisis is part of His (meta) divine End Times plan. Some premillennial Evangelicals interpret the environmental apocalypse as a hoax while others see environmental catastrophes as orchestrated by a vengeful God, and still others view it as a "sign" of the End Times and therefore a path to salvation. We explore these three variations below in relation to their nuanced depoliticizing and post-political interpretations.

Ecological catastrophes are seen by some Evangelicals as expressions of God's vengeance. Almost any environmental catastrophe is explained by some (largely American) Evangelical pastors as being the result of punishing the United States for some foreign policy decision or a perceived decline in American values. As an example of American foreign policy, a North Carolina pastor saw Hurricane Katrina as "God's judgment for sins as diverse as the US government's sanction of Gaza's evacuation and New Orleans' 'decadent night life'" (Bakker 2005, p. 798; Pearson 2006, p. 9). Such conclusions about increased destructive weather patterns are depoliticizing on the one hand by denying climate change had any influence, and on the other, that such weather patterns are caused not by human action, but by God (Barker and Bearce 2013).

Other Evangelical groups see climate change as a sign of the End Times and salvation but also see it as anthropogenic climate change by borrowing the warnings from climate scientists, hyperbolically interpreting it through biblical passages, especially the Book of Revelation. Two leading dispensationalist authors, Walvoord and Hitchcock (2007, p. 11), describe Revelation's environmental apocalypse as "acts of man":

> Acts of man, resulting in thousands of martyrs, and acts of God will combine to cause great disturbances in the world and universe. Stars will fall and planets will run off course, causing chaotic changes in climate.
>
> (Revelation 6:12–14; 16:8–9)

> Unnatural heat and cold, flooding, and other disasters will wipe out much of the food production of the world.
>
> (Revelation 6:6–8)

For Walvoord and Hitchcock, the consensus is that the worse our environmental problems become, the closer we are to being Raptured out of this world prior to the battle of Armageddon. These influential premillennialists have depoliticized climate change by suggesting both that the apocalypse is the gateway to Heaven and also that only God can change the world. To the American Christian Right, argues the religious studies scholar Michael Northcott (2004, p. 60), socio-economic and political environmental reform is an act of liberal Christianity that betrays the Gospel. Such groups then welcome the atrocities and inequalities of the world, consistent with neoliberal capitalism's meritorious-administrative-technological fix only transcribed here as a metaphysical fatalism. As stated by Garrard (2004, p. 8), "environmental crisis serves modern American conservative evangelists just as natural disasters served medieval millenarians: as a sign of the coming End, but not as a warning to avert it." Guyer (2007) argues that fundamentalist Evangelicals, in tandem with current Western capitalism, are obsessed with a determined, ultimate, and distant future at the expense of a "near future" defined by Fredrick Cooper (2005, p. 149) as a promise of "choice, political organization, responsibility, and accountability." In other words, a new temporality of depoliticization is promoted in which debate is lost to a future fatalism where time is predetermined to dispensations of truths about what is to come (Sturm 2021a; Sturm 2021b). Premillennialists then believe that humans should make no concerted effort to environmental betterment because they do not see their co-existence with the environment as inseparably entangled with it.

Many eco-luminaries promote a variety of depoliticizing apocalyptics that conclude the Earth has the agency to get revenge on humans or that humans are too far outside nature to live in harmony on Earth. Since Rachel Carson's popular book, *Silent Spring* (1962), it has become somewhat normalized among eco-activists to speak of environmental degradation in apocalyptic terms like "destruction" and "war on nature." Carson has not gone so far as to depoliticize climate change. Others have, however, by positing that the Earth is a living being or deity that has agency to end human life on Earth. For example, in chapter 8 of *Gaia* (1991), entitled "The People Plague," James Lovelock writes of humans as biblical petulance and has led a movement of people who would prefer humans disappeared so the planet could be natural once again (Pepper 2005, p. 3). Similarly the desire of the deep ecologists and groups like *Earth First!* to "return to nature," as Michael Watts put it, has a long pedigree in Western thought of "apocalyptic thinking that sees nature striking back" (Lee 1995; Watts 2005, p. 154).

More recently, the "Half Earth" project led by conservation scientists (Noss et al. 2012; Wilson 2016; Dinerstein et al. 2020) argues that half of the Earth should be set aside in a series of interconnected spaces for nature conservation as a global safety net. What on the surface reflects a radical proposal, the clearest description of the project by Wilson (2016, p. 3) has all of the trapping of a post-political argument: "only by setting aside half the planet in reserve, or more, can we save the living part of the environment and achieve the stabilization required for our own survival." Not to put Half Earth on equal footing, but many movements can be depoliticizing while being radical from the Trump movement to the KKK; crucially the processes of politicization in the direction of an equal, free, and solidary-based configuration are absent within the techno-managerialism of Half Earth. Here, truth and nature are co-constructional, thus legitimating ontological truth claims about nature and the certainty of apocalyptic aversion (Forsyth 2003, p. 4). These theories, sometimes referred to as "radical," of an inevitably vengeful Earth, in fact, depoliticize apocalyptic environmentalism by foregoing reformism through suggesting that humans are ephemeral beings that have removed themselves too far from nature to sustainably live on the planet. Anti-Gaia discourse has become even more apocalyptic, however, speaking instead of the "Medea Hypothesis" where instead of complex life being in symbiosis, microbial life—COVID-19 included—has worked together to create mass extinctions that are part of natural, and therefore depoliticized, cycles (Ward 2007).

Popular American academics and American politicians also provide their own variant of depoliticizing apocalyptic environmental discourses. In contrast to Swyngedouw's assumption that there exists a political consensus on climate change, in some places there is a lack of a consensus on climate change and needed action, which does not necessarily indicate a real democratic debate, but can be an indication of multiple depoliticizing strategies not just one built on consensus to technocratic action. Below we attempt to variegate apocalyptic positions, echoing McCarthy (2013, p. 23) who "urge[s] more modest and consciously situated claims regarding that state of contemporary 'politics writ large,'' especially in the United States. In the United States, for example, academics and politicians still debate this issue vigorously, but both leftward and rightward are often depoliticizing in different ways; thus, Swyngedouw does not capture the full range of post-political and depoliticizing maneuvers in his limited case studies. Between the right-wing politicians and the left-wing popular writers, there is a lack of consensus on climate change but a convergent depoliticization that does not question capitalism as the root of our environmental crisis. Among policy makers, US Senator Marco Rubio, for example, believes in catastrophic climate change but denied a human and thus industrial capital cause concluding, "I do not believe that the laws that they [climate scientists] propose we pass will do anything about it [climate change], except it will destroy our economy" (Bennett 2014). Senator Inhofe (2003), a conservative Oklahoma-based Evangelical and former Chair of the Senate Environment and Public Works Committee (SEPWC), has gone so far as to call climate change itself a "hoax." Rubio has naturalized the threat seemingly to maintain a neoliberal status quo. Indeed climate change skeptics, like free-market think tanks, energy lobbyists, industry advocates, and senators connected to this complex, stand to gain from a North-West passage or drill-bits in the Arctic. Other policy makers like the former Chair of the SEPWC, Barbara Boxer, used apocalyptic rhetoric in statements like "time is running out" concerning global warming which both loosely date-sets and defers to the future environmental issues through apocalyptic "psychological imminence" (Young 2006; McGinn 1998, p. 378). Boxer, despite being on the left and acknowledging the threat, has depoliticized climate change by suggesting it is a threat to humanity's immanent temporality, but like her European counterparts, stopping short of a fundamental and necessary critique of capitalism, instead choosing to work within the politico-economic tools of the system by offering technocratic fixes. The contrast between the policy makers illustrates a split or variegated apocalyptic consensus in the United States (Boykoff 2007).

The scientific community also uses this hyperbolic apocalyptic language, like "critical thresholds" (Rockström et al. 2009), "Omega Point," and "point of no return" (Alley 2004, p. 65), which in many ways echoes the threshold of an apocalyptic limit after which there is not simply change, as in systems-theory terms like Schelling's (1960) "tipping point," but rather a fundamental warning to avert disaster. Many scientists and popular scholars, however, often fail to politicize their findings beyond the present horizon. Naomi Klein (2013) summarizes another depoliticizing position among climate scientists succinctly, "in order to appear reasonable within neoliberal economic circles, scientists have been dramatically soft-peddling the implications of their research." She quotes Anderson and Bows' relevant 2012 *Nature Climate Change* article,

> in developing emission scenarios scientists repeatedly and severely underplay the implications of their analyses. When it comes to avoiding a 2°C rise, 'impossible' is translated into 'difficult but doable,' whereas 'urgent and radical' emerge as 'challenging' – all to appease the god of economics (or, more precisely, finance)... so geoengineering is increasingly proposed to ensure that the diktat of economists remains unquestioned.

The immediacy of the apocalyptic effects of climate change has the effect of persuasion and simultaneously instilling anxiety and fear, but some climate scientists want to limit the fallout related to the apocalyptic potentiality of their findings. Moreover, and similar to Swyngedouw's point, climate scientists' political efficacy is further muddied as Pilkey and Pilkey (2006) point out persuasively, where their emphasis on mathematical models is misleading to a democratic public because of thick jargon and an unwillingness to engage in the political dimensions of environmental threats.

The nuance and variety in and between secular and religious apocalyptic visions is largely lost in Swynegdouw's arguments of both depoliticization and the post-political. Above we extend and modify such environmental apocalyptic discourses beyond the consensus that Swyngedouw allows for. Without a nuanced and evidence-based variance of environmental apocalypses, the alliance-based democratic, contestatory, and radical transformative potential of the apocalypse is lost, and so too is a prescriptive millennialism. We explore below contestatory apocalyptic discourses that we argue have the potential to create unlikely alliances that are neither post-political nor neoliberal.

Contestatory

The conceptual polarization between the post-political, post-democratic consensus and the proper, authentic, real politics of *the* political (identified as evental, transformative ruptures), a polarization that frames much of Swyngedouw's discussion of apocalyptic discourses (even while he at times recognizes the ineliminable admixture of these domains), has the unfortunate consequence of overlooking or diminishing the importance, implications, and politicization possibilities of, what we will call here, contestatory apocalypses.

For the crafting of this term, and as guide for the discussion that follows, we are drawing upon the work of Leitner et al. (1007) in the collection *Contesting Neoliberalism: Urban Frontiers*. This important collection of essays makes a number of significant critiques of prominent depictions of neoliberalism, a set of critiques we believe have parallel implications for the emerging discourses that center on the post-political. Leitner et al. warn against top-down, neoliberal-centric analyses that can "reinforce its hegemonic status," "reify its ubiquity and power," and reproduce the "stereotypical images" promoted by its proponents (Leitner et al. 2007, pp. 2, 5, 11). Such perspectives fail to recognize how neoliberalism is constituted, reshaped, and exceeded by its ubiquitous "rival contestants" (Leitner et al. 2007, pp. 9–10). We argue the depiction of neoliberalism, the post-political, and the post-political environmental consensus as described in a series of essays by Swyngedouw under appreciates "these multiple and complex contestations," including alternative ecological visions, that are, indeed, capable of reinforcing and supplementing neoliberalism (or even helping it evolve toward even worse outcomes), but also of challenging, weakening, and ultimately possibly supplanting neoliberalism.

While understanding the post-political as a subversion of politics to the "art of the possible" and not a disavowing of alternative voices, we insist that the post-political is not an achieved and dominant condition, but rather as an emerging and perhaps increasingly predominant strategy and tendency of depoliticization (an evacuation of politics) by current regimes of power. We instead pay much needed attention to the multiple ways in which neoliberalism's mobilizing of depoliticizing apocalyptic discourses has been incessantly contested and challenged by alternative visions and organizations. What follows below is a selective inventory of the proliferation of diverse discourses exemplifying forms of organization, innovation, and inspiration that, while perhaps not the radicalized evental politics both we and Swyngedouw believe are needed, and in some cases are illustrative of desires for a darkening of our current conditions, nevertheless indicate "the eruption and persistence of [apocalypse-driven] challenges to neoliberal [and post-political] rule" (Leitner et al. 2007, p. 326). Echoing a critique made by Adrian Johnston on similar distinctions within the work of Badiou: divides between depoliticized situations and radical political events overlooks "how forms of concrete engaged praxis" might prefigure, precipitate, and participate in interruption, displacements, and reorganizations of dominant parameters (Johnston 2008, pp. 7, 13, 20, 2009, 48).

In this more contestatory form of apocalyptics, humans do have a role and possibly can take effort to change or reverse the direction of climate change outside often technocratic possibilities presented by capitalism, state governments, and elites. Instead both rightward and leftward politicizations that are not populist, what we call "contestatory," fit between the either/or made by Swyngedouw between post-political administrative and radical eventual ruptures of environmental apocalypticism. What follows below is a selective inventory of the proliferation of diverse discourses exemplifying contestatory forms of organization, innovation, and inspiration.

Contestatory discourses do not always take place from the political left, and authors advocating on behalf of the left make prescriptions that can be interpreted as rightward and include increased military powers, reduced governments, or authoritarianism that is divisive and not populist. For example, the popular scientist, James Lovelock, in a 2010 interview with the *Guardian* advocated anti-democratic and authoritarian rule to prevent an environmental apocalypse. Lovelock stated plainly,

> we need a more authoritative world.... we've become a sort of cheeky, egalitarian world where everyone can have their say. It's all very well, but there are certain circumstances – a war is a typical example – where you can't do that.

He concluded, "it may be necessary to put democracy on hold for a while" (Hickman 2010).

Among more leftward political eco-activists, both Green Parties and Greenpeace are contestatory organizations with aggressive reformist demands that are not necessarily supporting neoliberalism. Greenpeace takes up the Quaker practice of global witness in an attempt to change people's perceptions of human impact in the natural world, and so "Greenpeace actions have thus attacked specific forms of industrial production that threaten the reproduction of life" (Merchant 2005, p. 186). But Greenpeace is a light green reformist group that does well to "identify the foundational dynamics of environmental problems" but does little to "work to change them." As Wapner (1995, p. 300) summarizes the more radical enviromentalists' critique of Greenpeace, "advertisement does not affect or change the structures of power which create and support environmental destruction." Similarly, North American Green parties have taken an apocalyptic urgency and advocated political "reformism" to mitigate anthropogenic climate change. They have largely rejected any neoliberal technocratic or government-based solution consensus and instead "embraced principles such as anti-capitalism, social ecology, and women's, gay, and lesbian liberation" and "called for an independent radical politics outside the Democratic and Republican parties and promoted an ecologically-oriented cooperative commonwealth based on decentralized, democratic, public ownership of poverty and guaranteed housing, healthcare, and employment" (Merchant 2005, p. 179). Despite this radical potential, North American and European Green parties have to some extent become "professional-electorate" parties (Frankland et al. 2009). Green parties are calling for major transformations but are not fundamentally against government consensus and are generally comfortable with eco-friendly capitalism, suggesting instead that the system can be incrementally altered from within through party politics and more localized capitalisms.

For the depoliticized apocalyptic discourses explored in the last section, and to paraphrase Jameson (1994), it might be easier to think of the end of the world than think of incremental change in capitalism. Among social theorists, Goldstein's (2013) critique of Swyngedouw suggests just such an incremental politics, what they call, "appropriate technocracies" or alternative capitalisms where a Green New Deal can act as a constituency of change that increases demands on the system. Goldstein's critique suggests that a leftward contestatory that is non-neoliberal capitalist can be an incremental step toward a set of radical environmentalism alliances:

> building a radical environmental subjectivity, a popular and popularly radical environmentalism, may need to engage with the discourses and desires reflected in calls for green capitalism. We must be able to rework these discourses... while recognizing, respecting and amplifying the radical, liberatory aspirations that likely draw readers into these texts in the first place.
>
> (Goldstein 2013, p. 32)

What we would like to see, then, is a more eventual apocalyptic through alliances. We believe that the contestatory identity politics explored above can be radicalized, thus building on successes like a Green New Deal, for further transformation.

According to Parenti (2013, 2014), even those who desire radical social transformations must recognize that the immense scale and compressed timeframe of a variety of impending ecological catastrophes means that even radical movements should endorse realistic, short-term, reformist interventions into contemporary conditions and systems. Rather than expecting or demanding socialist transformations to precede solutions to impending climate chaos, reformist measures toward mitigation and adaptation should be pursued in order to "buy time," including such policies as a Carbon Tax, greater empowerment of the EPA, and redirecting the "shadow socialism" of government procurement toward green energy, technology, and goods (Parenti 2013). Such reformist interventions, argues Parenti, are not at odds with longer-term radical change, but can, rather, be seen as preconditions for systematic transformations as well as immediate opportunities for rethinking the role of the state, shifting the balance of power between classes, and reviving historically productive relations between reform and rebellion.

Such incremental democratic and pluralistic politics may provide potential for new spaces of resistance. Such political alliances can contribute to the transformative eventual potential of apocalyptic thinking. In other words, rather than majority forms of democracy, we advocate a more radical affinity-based set of relations across creedal differences in response to climate change apocalypses. William E. Connolly writes this, for example, will require including a "section of evangelicals into a larger movement made up of monotheists and nontheists of multiple types, whose affinities of spirituality draw them toward pluralism, egalitarianism and care for the earth" (Connolly 2010, p. xiii). Thus, this post-secular condition requires our acknowledged creedal affinities despite an assumption of secularity to open up intersubjectivity. What is missing, then, in the politicized views of climate change, is the everyday and localized politics that acknowledges the variegated, localized, and regional effects of climate change and various scales of alliances that can come together not simply for radical ends, but for democratic debate that acknowledges democracy is never rational or public. In the below section, we explore the radical, evental, and transformative potential from various groups. We advocate alliances that create potential to change one another rather than contestatory pluralist notions that are all too often limited to their own identity-based interests.

Radical

For Swyngedouw, a "radical socio-environmental political programme" transforms questions of climate change into questions and demands for greater democracy, insists on "a radical repoliticization of the 'economic,'" and the development of an oppositional proliferation of socio-environmental futures, which will hopefully lead to more egalitarian socio-metabolic processes and a more equitable distribution of social power (Swyngedouw 2010a, p. 204). While we consider ourselves in broad agreement with his call for a radical democratic movement that pursues the "emancipatory possibilities of apocalyptic life" in the contemporary conditions of the already unfolding, combined, and uneven apocalypse (Swyngedouw 2013, p. 17), we think such a movement needs to include a recognition, engagement with, and enhancement of already existing radical discourses, particularly those that are mobilizing visions of the apocalypse to frame fundamental critiques of capitalism and spur the emergence of a radical movement. What is needed is less a repoliticization of the economic, but rather a recognition of its always already contested operation and the existence of a myriad of radical apocalyptic discourses critiquing core features of capitalism and calling for revolutionary change by a variety of eco-activists, scientists, religious figures, and social theorists. Rather than a repoliticization, what is needed is an encouragement of those critiques already taking place and, in line with Swyngedouw's own call, an increased proliferation of discourses that use the unfolding apocalypse as an opportunity to inspire, connect, and organize democratic mobilizations. A continued, varied, and intensified radical environmental apocalyptics can help expand contemporary contestations and catalyze the emergence of what we will call, in a synthetic and slightly askew reading of several discourses below, radical alliances among the eco-precariat.

Swyngedouw is not alone in his assessment of apocalyptic fervor and rhetoric furthering post-political ends. The theologian Catherine Keller (1999, p. 52) suggested as much of environmental apocalypticism:

> Of course apocalypticism itself, whether in religious or secular rhetoric, tends to overheat and to burn us out; or to provoke modes of denial which invite us to chill out and not worry. The result of either extreme seems to be precisely the lukewarmth of indifference, which cooperates with corporate interests.

Furthermore, climate change apocalyptic rhetoric has been adopted, according to the religious studies scholar Michael Northcott (2008), to create an apocalyptic politics of fear by governments and elites to further otherwise unsavory political and economic ends. Making demands on the system through fear and anxiety opens up a conversation about the potential for rechanneling the energies of apocalyptic toward a more radicalized set of everyday political changes. As Kovel (2007, p. 5) summarizes these religious voices as illustrating the repoliticization and emancipatory potential of the apocalypse, "how can the environmental changes ahead that will preserve and even enhance life be conceived unless we recover the imaginative power of apocalypse?" As Cohn (1957) has classically paralleled in his book, *The Pursuit of the Millennium*, "mystical anarchist" millennial movements can be profoundly revolutionary against oppression and persecution. Revelation can be read as a promotion of radical upheaval, resistance, and transformative eventalism. The use of metaphors and symbols, not least Rome, is promoted by some eco-theologians as transgressive of capitalist-led climate change.

Evangelicals of the more radical theological persuasion view environmental apocalypticism within its poetic original meaning to unveil new ways of thinking about how we live in this world. Revelation was a symbolic resistance to capitalism and/or empire (Megoran 2013) bound up with corruption of nature. Keller's (2004) counter-apocalypse both enlists the anti-imperial elements of Revelation and is counter to the escapist, fatalistic, and thus post-political fundamentalist Evangelical definitions of environmental catastrophe. Keller (1999, pp. 57–58) writes in relation to apocalyptic climate change visions:

> Apocalypse means unveiling, disclosure. So we had better tease its contemporary incarnations out of their bitter sense of closure, toward their own dis/closive potential. Counter/apocalypse allows us to strengthen the relation of the apocalyptic unconditional to the relativizing, relational conditionals of the larger biblical tradition... in the process perhaps emergency gives way to emergence, and uncertainty to adventure.

Similarly, not all climate scientists have depoliticized or shirked to the technocratic-administrative post-political solution to climate change. Climate scientists have also politicized their findings, albeit in apocalyptic rhetoric. Answering the question, "Is Earth Fucked?" Brad Werner a geophysicist at the Fall Meeting of the American Geophysical Union replied, "if we're thinking about the future of the earth, and the future of our coupling to the environment, we have to include [human] resistance [to capitalism] as part of that dynamics... [that is] really a geophysics problem" (Klein 2013, np). Werner's position, among a growing group of other scientists like Grantham (2012), is attempting to re-democratize environmental apocalypses and even advocating working outside the law.

John Bellamy Foster and Brett Clark argue that the source of our rapidly worsening environmental conditions is the process of capitalist accumulation and any adequate solution to the impending global environmental catastrophe requires a "wholesale redirection of the economy" that includes an overcoming of the "antisocial results of private ownership," a revolutionary transition to more rational forms of social planning, a decisive downshift toward more sustainable levels of production, and reorientation toward human community rather than rampant accumulation (Foster and Clark 2012, pp. 5 and 12–13). Foster and Clark (2012, pp. 2 and 12) note, humanity is confronted by its ultimate crisis: not merely climate change, but a devastating sequence of "planetary rifts" including ocean acidification, ozone layer destruction, fresh water shortages, land-cover changes, soil degradation, and species extinction. What is needed is the emergence of a radical "environmental proletariat" capable of confronting and transforming the fundamental nature of our current social system (Foster and Clark 2012, pp. 10, 12).

The most important radical theorist to intervene in this debate and help craft cognitive maps of the revolutionary potential within apocalyptic discourses and events is Slavoj Žižek. What is most striking in Žižek's analysis of apocalyptic discourses, probabilities, and politics is how he strives to recode the apocalypse within a frame of proletarian revolution, but one dramatically different from orthodox Marxism. In Žižek's communistic apocalyptic thought, the proletarian political revolution capable of overthrowing capitalism once again becomes possible ("the impossible is becoming possible"), but now through the conjuncture of several apocalyptic trends that Žižek (2011, p. 328) sees as intertwining together into simultaneous ruptures of contemporary capitalism. Drawing upon the work of Ed Ayers and Holmes Rolston, Žižek (2009, p. 93) identifies four ecologically problematic "spikes" of accelerated development: rapid growths in population, the depletion of vital yet finite natural resources, escalating carbon emissions, and the sudden mass extinctions of species. In addition, these ecological trends are connected to several other antagonisms plaguing capitalism, including private property's problematic enclosures of "intellectual property," life-altering enhancements of techno-science (in particular biogenetics), and intensifications of social inequalities and geographical apartheids worldwide (Žižek 2009, p. 91). That each crisis is deepening along with each of the others indicates a considerable, growing, and perhaps existential management crisis for contemporary capitalism. Instead of doom and gloom, Žižek sees an opening: as in much of revolutionary history, and perhaps soon in similar apocalyptic conditions, "Revolutionaries have to wait patiently for the (usually very brief) moment when the system openly malfunctions or collapses, have to exploit the window of opportunity, to seize power…." (Žižek 2009, p. 90). Žižek's post-Apocalyptic millennium.

As Žižek notes, one of the great analytical and organizational deadlocks of Western Marxism was the trouble identifying a revolutionary agent, or explaining the troubling absence of one (Žižek 2009, pp. 88–89). "Western Marxism was… engaged in a constant search for other social agents who could play the role of the revolutionary subject, as understudies who might replace the indisposed working class: Third World peasants, students, intellectuals, the excluded…". It is this final term, the Excluded, that Žižek argues now serves as the revolutionary focal point, though not the sole subject. The Excluded refers to the marginalized, the forgotten, the dispossessed—not a proletariat, but many. The conjunctural nature of the apocalypse indicates a diversity of "processes of proletarianization," including widespread dispossession of whole people's biology and knowledge. Such a diversification of immiserations is capable of spurring multiple revolutionary enthusiasms and means that "a new emancipatory politics will stem no longer from a particular social agent, but from an explosive combination of different agents" (Žižek 2009, p. 92). It is the subversive sewing together of these variegated fights that makes a radical politics of apocalypse possible. Indeed, "the key to the ecological crisis does not reside in ecology as such," but rather resides in how each of the different antagonisms that are definitive for our times and immediate future is incorporated into demands for universal justice (Žižek 2011, p. 334).

In Žižek's (2011, p. 352) most surprising rhetorical move in his radical politicization of the apocalypse, he draws attention to the formal similarities of a radical emancipatory logic and the "ridiculous" and "dangerous" Christian fundamentalist apocalypticism. Both conceive of the apocalypse as a question of justice and redemption and both demand kinds of unconditional commitment from their supporters. Yet, Žižek's selective recuperation of Christian apocalyptics is not without significant qualifications: any revolutionary movement oriented toward the apocalypse must embrace ecology and contingency. He notes that even with these parallels in formalism and fidelity between Christians and radicals, a radical apocalyptic politics must bring plans for a proletarian upheaval "into closer contact with secular ecologism, thereby conceiving the threat of annihilation as the chance for a radical emancipatory renewal" (2009, p. 94). Chance is key here. Žižek (2011, p. 350) is emphatically not endorsing a communist teleology through the backdoor of the apocalypse. For him, radical apocalyptic movements must recognize the "utter contingency of nature." Certainty, whether objectively scientific or religiously metaphysical, is rejected in favor of an "openness to radical contingency" of "new possibilities" and their perhaps "unexpectedly positive results" (Žižek 2011, pp. 351–352).

Conclusion

Our currently unfolding ecological breakdown calls not only for immediate political and policy interventions, but also the emergence of an emancipatory movement capable of crafting alliances with progressive contestatory movements, geographically dispersed forms of discontent and resistance, divergent and antagonistic positions, and radicalizing the movement as a whole toward post-capitalist politics and ecologies. The most intriguing possibility for us is the emergence of, in a twist on Foster and Clark's (2012, p. 16) call for an "environmental proletariat" and contemporary discourses around precarious labor and "human-environmental precarities" (Morrissey 2020, p. 1), a global eco-precariat as an emancipatory subject. The eco-precariat can be understood as a political signifier that stitches together the political fights of all those rendered precarious by the global capitalist economy and all those that rightly recognize their lives as rendered precarious by impending ecological catastrophe. Žižek's identification of the masses of slum dwellers as a potential contemporary emancipatory subject is then prescient and precise, as such masses occupy the harshest intersection of capitalist and climatic destruction. Yet the politics to come of the eco-precariat is fundamentally open to all, like an ecological solidarity movement, a combinatory linkage across divergent and antagonistic positions for the sake of immediate demands and the provocation of the impossible events of emancipation. As we argue in this chapter, Swyngedouw's analysis of post-political apocalyptic discourses may simplify the variety and differences of depoliticizing apocalyptic discourses, overlook the existence and significance of ongoing contestations, and fail to explore the diverse radical apocalyptic discourses that continue to be mobilized against capitalism and its ecological consequences; he nevertheless is right to pinpoint apocalyptic conditions and discourses as a critical terrain of contemporary emancipatory struggles: "political choices have to be made and sides taken," so let us "fully endorse the emancipatory possibilities of apocalyptic life" (Swyngedouw 2013, p. 17). Such an end can be a new millennium (Sturm 2021a).

In an effort to build upon the elements of this radical apocalypticism reviewed in this chapter, prompted by Swyngedouw (2010b), and as provocation to debate as well as a guide for future work, we conclude this chapter with a brief summary of the six principles toward a manifesto for building radical apocalyptic politics *for* the Anthropocene. We consider these principles as the core for new radical apocalyptic discourses.

1 *Earthly.* A progressive apocalyptics must reject any transcendental ideas, claims, or expectations: there is no God, no Savior, no teleological forces determining the shape and impact of the apocalypse. The apocalypse is above all an earthly event. The matter and forces of the earth alone must be seen as the causes, conditions, consequences, and responses of apocalypse. We must insist on a materialist perspective. The stakes are not souls, salvation, heaven, and metaphysical purpose; but bodies, brains, pleasure, projects, and earthly justice.
2 *Immanent.* However imminent the most catastrophic events may be, the apocalypse is profoundly immanent to our present situation. The apocalypse is ahead of us, but also upon us, behind us, and within us. The forces contributing to impending ecological, economic, and social catastrophe already operate in and profoundly shape the present. Recognition of this immanence can sharpen our sense of complicity, responsibility, and potential.
3 *Contingency.* We must refuse any claims of access to divine knowledge, or perfectly clairvoyant prophecy, or absolutely unassailable objective truths. Our materialism is grounded in a radical contingency (i.e. a foundational unpredictability that means things could have been different, and can always be different) that forces us to consider our apocalypticism as fundamentally grounded in transcendentally ungroundable *decisions.* We accept being uncertain. Our positions and actions are pragmatic preparations and provocations, not predetermined participation in already scripted revelation.

4 *Conjunctural.* The apocalypse includes political, economic, technological, and ecological dimensions and causes. An appreciation of these multiple dimensions of a composite apocalypse is crucial for considering the contingency of the unfolding of any apocalypse, the diverse causal pathways of its production, the troubling compounding and amplification of apocalyptic traumas, and the opportunity, indeed the *need*, for a large-scale, multi-dimensional, socio-ecological transformation.
5 *Uneven.* Different places and different people are impacted and will be impacted differently what Williams (2010, p. 11) refers to as the "combined and uneven apocalypse." The poor—the eco-precariat—suffer the most. The apocalypse is overcoded by class divisions and inequality. A radical apocalypticism intent on earthly justice requires an awareness and emphasis on the geographical and social unevenness in the causes and consequences of the apocalypse. Our expectations, plans, and demands must reflect this awareness.
6 *Redemptive.* The apocalypse contains unprecedented transformative potential. All efforts to reduce the catastrophe to an apocalypse, to an event locatable (and thus solvable) within the coordinates of present systems of power and profit should be subverted. The apocalypse is an opportunity to transform everything. Perhaps a certain kind of apocalyptics is our new utopianism—in the sense of providing an orientation and path toward a future revolutionary upheaval and ultimate liberation. We should see the apocalypse as a possibility for redemption: of the countless horrors of human history and of our complicity in any of the causes of an apocalyptic event (we need to be redeemed, all of us, though some more than others).

References

Alley, R. (2004) Abrupt climate change. *Scientific America* (Nov), pp.62–69.

Anderson, K. and Bows, A. (2012) A new paradigm for climate change. *Nature Climate Change* 2, pp.639–40.

Bakker, K. (2005) Katrina: The public transcript of disaster. *Environment and Planning D: Society and Space* 23(6), pp.795–809.

Barker, D. and Bearce, D. (2013) End-times theology, the shadow of the future, and public resistance to addressing global climate change. *Political Research Quarterly* 66(2), pp.267–279.

Bennett, B. (2014) Marco Rubio says human activity isn't causing climate change. *Los Angeles Times* 11 March. www.latimes.com/nation/politics/politicsnow/la-pn-rubio-denies-climate-change-20140511-story.html (last accessed 13 May 2019).

Boykoff, M.T. (2007) Flogging a dead norm? Newspaper coverage of anthropogenic climate change in the United States and United Kingdom from 2003 to 2006. *Area* 39(4), pp.470–481.

Brenner, N., Peck, J. and Theodore, N. (2010) Variegated neoliberalization: Geographies, modalities, pathways. *Global Networks* 10(2), pp.182–222.

Carson, R. (1962) *Silent spring*. Boston: Houghton Mifflin.

Cohn, N. (1957) *The pursuit of the millennium*. London: Secker and Warburg.

Connolly, W. E (2010) Evangelism, secularism and pluralist possibility. In J. Dittmer and T. Sturm eds. *Mapping the end times: American evangelical geopolitics and apocalyptic visions*. Farnham: Ashgate. pp. xi–xvi.

Cooper, F. (2005) *Colonialism in question: Theory, knowledge, history*. Berkeley: University of California Press.

Cunningham A, and Grell, O. P. (2000) *The four horsemen of the Apocalypse: Religion, war, famine and death in reformation Europe*. Cambridge: Cambridge University Press.

Darling, J. (2014) Asylum and the post-political: Domopolitics, depoliticisation and acts of citizenship. *Antipode* 46(1), pp.72–91.

Diken, B. (2014) Religious antinomies of post-politics. In J. Wilson and E. Swyngedouw eds. *The post-political and its discontents: Spaces of depoliticisation, spectres of radical politics*. Edinburgh: Edinburgh University Press. pp.126–145.

Dinerstein, A. R., Joshi, C., Vynne, A. T. L., Lee, F. Pharand-Deschênes, M., França, S., Fernando, T., Birch, K., Burkart, G. P. Asner, D. and Olson, A. (2020) Global Safety Net to reverse biodiversity loss and stabilize Earth's climate. *Science Advances* 6(36), pp.1–13.

Dittmer, J. and Sturm, T. eds. (2010) *Mapping the end times: American evangelical geopolitics and apocalyptic visions*. Farnham: Ashgate.

Forsyth, T. (2003) *Critical political ecology*. London: Routledge.

Foster, J. B. and Clark. B. (2012) The planetary emergency. *Monthly Review* 64(7), pp.1–25.

Frankland, G., Lucardie, P., and Rihoux, B. eds. (2009) *Green parties in transition: The end of grassroots democracy?* Farnham: Ashgate.

Garrard, G. (2004) *Ecocriticism*. London: Routledge.

Goldstein, J. (2013) Appropriate technologies? Green capitalist discourses and post capitalist desires. *Capitalism Nature Socialism* 24(1), pp.26–34.

Grantham, J. (2012) Be persuasive. Be brave. Be arrested (if necessary). *Nature* 491(7424), p.303.

Guyer, J. (2007) Prophecy and the near future: Thoughts on macroeconomic, evangelical and punctuated time. *American Ethnologist* 34(3), pp.409–21.

Harper, E. (2019) Ecological gentrification in response to apocalyptic narratives of climate change: The production of an immuno-political fantasy. *International Journal of Urban and Regional Research*. https://doi.org/10.1111/1468-2427.12842

Hickman, L. (2010) James Lovelock on the value of sceptics and why Copenhagen was doomed. *The Guardian*, 29 March. www.theguardian.com/environment/blog/2010/mar/29/james-lovelock (last accessed 1 May 2019).

Inhofe, J. (2003) 149 CONG. REC. S10012, 10022 (daily ed.). *Congress Recordings*. 28 July.

Jameson, F. (1994) *Seeds of time*. New York: Columbia University Press.

Johnston, A. (2008) *Žižek's ontology: A transcendental materialist theory of subjectivity*. Evanston: Northwestern University Press.

Johnston, A. (2009) *Badiou, Žižek, and political transformations: The cadence of change*. Evanston: Northwestern University Press.

Keller, C. (1999) The heat is on: Apocalyptic rhetoric and climate change. *Ecotheology* 7(1), pp.40–58.

Keller, C. (2004) *God and power: Counter-apocalyptic journeys*. Minneapolis: Fortress Press.

Kovel, J. (2007) Facing end-time. *Capitalism Nature Socialism* 18(2), pp.1–6.
Klein, N. (2013) How science is telling us all to revolt. *New Statesman* 29 October. www.newstatesman.com/2013/10/science-says-revolt (last accessed 1 May 2019).
Lee, M. (1995) *Earth first! Environmental apocalypse*. New York: Syracuse University Press.
Leitner, H., Peck, J., and Sheppard, E. (2007) *Contesting neoliberalism: Urban frontiers*. New York: The Guilford Press.
Lovelock, J. (1991) *Gaia: The practical science of planetary medicine*. London: Gaia.
Lynch, R., Sturm, T., and Webster, J. (2021) Apocalyptic futures: Morality, health and wellbeing at the end of the world. *Anthropology & Medicine* 28(1), pp.1–12.
McCarthy, J. (2013) We have never been 'post-political.' *Capitalism Nature Socialism*, 24(1), pp.19–25.
McCammack, B. (2008) Hot damned America: Evangelicalism and the climate change policy debate. *American Quarterly*, 59(3), pp.645–668.
McGinn, B. (1998) The last judgment in Christian tradition. In B. McGinn ed. *Apocalypticism in western history and culture, Vol. 2*. New York: Continuum. pp.394–395.
Megoran, N. (2013) Radical politics and the Apocalypse: Activist readings of Revelation. *Area* 45(2), pp.141–147.
Merchant, C. (2005) *Radical ecology: The search for a livable world*. New York: Routledge.
Morrissey, J. (2020) Mobilising the language of emergency: Human security and climate action discourse. *Irish Studies in International Affairs* 31(1), pp.1–12. https://doi.org/10.3318/ISIA.2020.31.11
Mouffe, C. (2005) *On the political*. London: Routledge.
Northcott, M. (2004) *An angel directs the storm: Apocalyptic religion and American empire*. London: I.B. Tauris.
Northcott, M. (2008) *A moral climate: The ethics of global warming*. London: Darton, Longman, and Todd.
Noss, R.F. et al. (2012) Bolder thinking for conservation. *Conservation Biology* 26, pp.1–4.
Parenti, C. (2013) A radical approach to the climate crisis. *Dissent* Summer.
Parenti, C. (2014) Climate change: What role for reform? *Monthly Review* 65(11), pp.49–51.
Pearson, S. (2006) *A brief history of the end of the world*. New York: Carroll and Graf.
Pepper, D. (2005) Utopianism and environmentalism. *Environmental Politics*, 14(1), pp.3–44.
Pilkey, O. and Pilkey-Jarvis, L. (2006) *Useless arithmetic: Why environmental Scientists can't predict the future*. New York: Columbia University Press.
Rancière, J. (2001) Ten Theses on Politics. *Theory & Event* 5(3), https://doi.org/10.1353/tae.2001.0028
Rockström, J., Steffen, W., Noone, K., and Persson, Å. (2009) A safe operating space for humanity. *Nature* 461, pp.472–75.
Schelling, T. (1960) *The strategy of conflict*. Cambridge: Harvard University Press.
Skrimshire, S. ed. (2010) *Future ethics: Climate change and apocalyptic imagination*. London: Continuum.
Skrimshire, S. (2013) Climate change and apocalyptic faith. *WIREs Climate Change* 5(2), pp.264–78.
Stephens, R. J. and Giberson. K. W. (2011) *The anointed: Evangelical truth in a secular age*. Cambridge: Belknap Press.

Sturm, T. (2021a) God's just Gaza war: Futurity foreclosed through evangelical apocalypse as orthogonal promise of Eretz Israel. *Geografiska Annaler: Series B, Human Geography*. https://doi.org/10.1080/04353684.2021.1891557

Sturm, T. (2021b) Hal Lindsey's geopolitical future: Towards a cartographic theory of anticipatory arrows. *Journal of Maps* 17(1), pp.39–45.

Sturm, T. and Albrecht, T. (2021) Constituent Covid-19 Apocalypses: Contagious conspiracism, 5G, and viral vaccinations. *Anthropology & Medicine* 28(1), pp.122–139.

Swyngedouw, E. (2010a) Apocalypse forever? Post-political populism and the spectre of climate change. *Theory Culture Society* 27(2–3), pp.213–232.

Swyngedouw, E. (2010b) The communist hypothesis and revolutionary capitalisms: Exploring the idea of communist geographies for the twenty-first century. *Antipode* 41(s1), pp.298–319.

Swyngedouw, E. (2013) Apocalypse now! Fear and doomsday pleasures. *Capitalism Nature Socialism* 24(1), pp.9–18.

Swyngedouw, E. (2017) Unlocking the mind-trap: Politicising urban theory and practice. *Urban Studies*, 54(1), pp.55–61.

Swyngedouw, E. and Ernstson, H. (2018) Interrupting the Anthropo-obScene: Immuno-biopolitics and depoliticizing more-than-human ontologies in the Anthropocene. *Theory, Culture & Society*, 35(6), pp.3–30

Walvoord, J. F. and Hitchcock, M. (2007) *Armageddon, oil, and terror*. Carol Stream: Tyndale.

Wapner, P. (1995) In defense of banner hangers: The dark green politics of Greenpeace. In B. R. Taylor ed. *Ecological resistance movements*. Albany: State University of New York Press. pp.300–314.

Ward, P. (2007) *Under a green sky: Global warming, the mass extinctions of the past, and what they can tell us about the future*. Washington: Smithsonian.

Watts, M. (2005) Nature: culture. In P. Cloke and R. Johnston eds. *Spaces of geographical thought: Deconstruction human geography's binaries*. Thousand Oaks: Sage. pp.142–174.

Williams, E. C. (2010) *Combined and uneven Apocalypse*. Park Lane: Zero Books.

Wilson, E.O. (2016) *Half-Earth: Our planet's fight for life*. London: Liveright Publishing.

Wilson, J. and Swyngedouw, E. (2014) Seeds of dystopia: Post-politics and the return of the political. In J. Wilson and E. Swyngedouw eds. *The post-political and its discontents: Spaces of depoliticisation, spectres of radical politics*. Edinburgh: Edinburgh University Press. pp.1–22.

Young, S. (2006) Boxer pledges shift on global warming policy with new Senate role. *Huffington Post* 9 November. www.huffingtonpost.com/2006/11/10/sen-boxernow-head-of-env_n_33824.html (last accessed 1 May 2019).

Žižek, S. (2009) *First as tragedy, then as farce*. London: Verso.

Žižek, S. (2011) *Living in the end times*. London: Verso.

Index

Note: *Italic* page numbers refer to figures. Page numbers followed by "n" denote endnotes.

www.ingramcontent.com/pod-product-compliance
Lightning Source LLC
LaVergne TN
LVHW021124300926
849160LV00010B/848